NARRATING SUSTAINABILITY THROUGH STORYTELLING

Stories and narratives are powerful tools for explaining the world around us. This book explores storytelling as a way of engaging audiences with sustainable development issues and reflects on the opportunities and limitations of storytelling for sustainability as an innovative approach to sustainability communication.

Bringing together voices and perspectives from research and practice, this volume explores the ways in which storytelling can support change toward sustainability. Unlike other anthologies, the book first provides a sound scientific basis by unfolding the storytelling approach and presenting empirical studies on its effects. It clarifies important terms and presents recent findings on the impact of storytelling on sustainability from an extensive 3-year research project on this question. The second part shows how storytelling can be used in different fields of practice to communicate sustainability in more engaging and effective ways. Here, the main focus is on not only case-based accounts of positive change but also tensions, arising from the application of storytelling for sustainability in journalism, higher education, corporate communication, or science communication.

Combining theory with practical examples, this innovative book will be a great resource for students and scholars of environmental communication and sustainable development, as well as professionals working in related fields.

Daniel Fischer is an associate professor of Consumer Communication and Sustainability at Wageningen University in the Netherlands and visiting professor of Sustainability Communication at the Leuphana University of Lüneburg, where he led the SusTelling project. From 2018 to 2020, he was an assistant professor at the School of Sustainability at Arizona State University, USA. Together with his research group SuCo2, short for Sustainable Consumption and Sustainability Communication (www.suco2.com), he researches how sustainable lifestyles can be promoted through communication and learning.

Sonja Fücker is a research associate at the Research Institute of Social Cohesion at the University of Hannover. Previously, she worked at the Institute for Environmental and Sustainability Communication (INFU) at the Leuphana University of Lüneburg. As a sociologist, her research focuses on the reception and impact of science communication. She is interested in how people acquire scientific knowledge in their everyday lives and what influence narrative and participatory transfer methods have on this.

Hanna Selm worked as a research associate at the Institute for Environmental and Sustainability Communication (INFU) at the Leuphana University of Lüneburg (Germany) and has been a member of the SuCo2 research group. As a cultural scientist, she has contributed to various (research) projects in the field of sustainability communication. She is particularly interested in researching and testing new formats of sustainability communication.

Anna Sundermann is pursuing a PhD at the Faculty of Sustainability at the Leuphana University of Lüneburg and working as a curriculum developer at Leuphana Graduate School. From 2013 to 2020, she conducted research on various projects at the Institute for Environmental and Sustainability Communication (INFU) at the Leuphana University of Lüneburg (Germany). In her PhD thesis, she brings her degree in psychology to bear on sustainability-related learning processes of students in higher education to find out whether the current integration of sustainability in higher education enables and motivates students to participate in societal change.

Routledge – SCORAI Studies in Sustainable Consumption

This series aims to advance conceptual and empirical contributions to this new and important field of study. For more information about The Sustainable Consumption Research and Action Initiative (SCORAI) and its activities please visit www.scorai.org.

Sustainable Lifestyles after Covid-19
Fabián Echegaray, Valerie Brachya, Philip J Vergragt and Lei Zhang

Sustainable Products in the Circular Economy
Impact on Business and Society
Edited by Magdalena Wojnarowska, Marek Ćwiklicki and Carlo Ingrao

Narrating Sustainability through Storytelling
Edited by Daniel Fischer, Sonja Fücker, Hanna Selm and Anna Sundermann

The Low-Carbon Good Life
Jules Pretty

For more information about this series, please visit: www.routledge.com/Routledge-SCORAI-Studies-in-Sustainable-Consumption/book-series/RSSC

NARRATING SUSTAINABILITY THROUGH STORYTELLING

Edited by Daniel Fischer, Sonja Fücker, Hanna Selm, and Anna Sundermann

sponsored by

DBU

Deutsche
Bundesstiftung Umwelt

www.dbu.de

Routledge
Taylor & Francis Group
LONDON AND NEW YORK

earthscan
from Routledge

Cover image: Saskia Ostner

The project has been funded and sponsored by Deutsche Bundesstiftung Umwelt (DBU)

First published 2023
by Routledge
4 Park Square, Milton Park, Abingdon, Oxon OX14 4RN

and by Routledge
605 Third Avenue, New York, NY 10158

Routledge is an imprint of the Taylor & Francis Group, an informa business

British Library Cataloguing-in-Publication Data
A catalogue record for this book is available from the British Library

ISBN: 978-1-032-35269-5 (hbk)
ISBN: 978-1-032-35270-1 (pbk)
ISBN: 978-1-003-32614-4 (ebk)

DOI: 10.4324/9781003326144

Typeset in Bembo
by Apex CoVantage, LLC

CONTENTS

FOREWORD

Sustainability Communication "Like Gold"?

When at the end of the 1980s teleshopping channels sprang up like mushrooms on television, I was one of those who every once in a while sat fascinated and at the same time stunned watching these nonstop commercials. With mixed feelings, I watched how everyday household items were being sold to women and men. The principle behind these teleshopping shows, so-called infomercials, works based on impulse purchasing, that is, buying products because of an instantaneous and spontaneous impulse. These impulse sales are triggered by product or packaging design as well as the shopping atmosphere (light, music, temperature, etc.).

The success of the advertising channel, so I thought at the time, depended less on the product itself and more on the presenter's ability to put what they were selling in the right light and build up a personal relationship with the potential customer. And this was quite skillfully done. The presenter talked about an approaching holiday, a meal with friends, planning a daughter's wedding celebration, a trip to visit the in-laws, or a granddaughter's concert at school. They were everyday situations many viewers could relate to – and one in which the gold chain being sold would play a key role. These stories were a link between the product and the personal lives of the viewers that – from the seller's point of view – would trigger an unreflective impulse and a quick decision to purchase the item.

It is obvious that this type of product advertisement does not usually address sustainability issues, and the principle of impulse buying cannot be easily transferred to this context – nor would it be desirable. Nevertheless, there are certainly similarities. Undoubtedly, a major challenge in sustainability communication is to reduce complexity to a level that laypersons can understand, without losing evidence-based disciplinary standards. At the same time, sustainability issues should be presented so that they arouse interest, are relevant to everyday life, and show as many concrete options for taking action as possible. Storytelling approaches have the potential of not only conveying information and explaining problems but also evoking emotions and thus creating an emotional closeness between the topic and the reader. This can lay an important foundation for changing behaviors. At the same time, oversimplification carries with it the danger of manipulation or misinformation. Here, it's crucial to strike the right balance between information and entertainment.

In journalistic practice today, we find a variety of different storytelling approaches that, at first glance, show a wide range of characteristics and effects. So far, however, there are only a few scientifically validated findings of the actual effects of these communication approaches. *Storytelling in Sustainability Communication: Perspectives and Evidence* is the first systematic research project into how narrative forms achieve their effects in sustainability contexts. The goal of this project then is not only to close a gap in communication research but also to make an important contribution to further professionalizing the use of narrative formats in sustainability communication and strengthening their effectiveness in practice.

This publication summarizes the findings obtained during a research project in a form that is suitable for traditional journalism as well as the fields of science and corporate communications. On the basis of empirical findings, the possible uses of storytelling, including its limitations, are explained, and how it achieves its effects. At the same time, the book provides a behind-the-scenes look at the experiences of communication experts, who share their perspectives on the practice of storytelling.

Unlike the infomercials described earlier, sustainability communication is not about triggering short-term buying decisions or unreflective changes in behavior. Authenticity and credibility, the creation of trust, identification with the topic, and options for taking action are – in addition to the dramaturgical structure and the use of entertaining elements – just a few of the many other aspects that make up a good story for advancing sustainability. If the focus in advertising is on buying a product, storytelling is more about the story behind the product, about, for example, how resources are used and under what social and ecological conditions this product was manufactured, how it was used, disposed of, and returned to the product life cycle, or how it contributes to reducing the burden on the environment. From an educational perspective, storytelling for sustainability is a serious effort to uncover sometimes very complex interrelationships and describe them in an understandable way, establish a connection to our own lives, and bring about a change in our behavior as consumers. The focus is not on the "gold chain" itself but rather on how the product impacts the environment and our thoughts and actions – this is the material that makes for a good story. The central question addressed in this book is how storytelling for sustainability can be used on both the content and the behavioral levels, as well as on the level of craftsmanship.

There were many reasons why the German Environmental Foundation funded this project. Its important contribution to research on the impact of storytelling in sustainability communication was an ideal complement to various other DBU funding projects that deal with issues such as constructive sustainability journalism or contribute to the training of often young journalists interested in sustainability. But it is the approach of presenting research findings – a new educational module, an online self-study unit, and this volume – that was especially persuasive. Fortunately, the project team was able to draw on the expertise of a renowned advisory board.

It is to be hoped that the journalistic format of storytelling will not only find broader use in sustainability communication but also that the approach – in a substantially more sophisticated way – will be just as successful at reaching their target groups as the concept of teleshopping channels mentioned earlier. As is so often the case, it is quality that counts, and this publication succeeds in doing just that. I should mention in closing that the completely overpriced "gold chain" was a chain with a "gold appearance", a piece of deceptive advertising that has no place in sustainability communication.

Thomas Pyhel
German Federal Environmental Foundation
Osnabrück, Germany
May 2022

CONTRIBUTORS

Jennifer Fritz is a storyteller and conceptual designer. She studied British and American Studies for her bachelor's degree and history for her master's degree at the University of Konstanz. Through her traineeship at Südwestrundfunk, a regional public broadcasting station, she discovered a love for digital media and writing. She believes that digital learning has the potential to be fun and change things at the same time. Her specialty is stories that communicate knowledge. As a freelancer, she works as an editor, writer, speaker, and instructional designer.

Daniel Hanss is a professor of environmental psychology and sustainability at Darmstadt University of Applied Sciences. He leads the specialization in environment and sustainability in the business psychology courses there. Since 2018, he has been building a citizen panel at Darmstadt University of Applied Sciences (https://buergerpanel.h-da.de), which allows people in the region to actively voice their opinions about innovation in transformation processes. In his research, he is interested in, among other things, what motivates people to contribute to sustainable development in their everyday lives and how scientific knowledge can contribute to this in transformative research and practice projects.

Sabine Hoscislawski is a research associate at the Institute for European Politics in Berlin. Previously, she was a trainee in the press and public relations department of the nonprofit organization for science communication Wissenschaft im Dialog (WiD). Before that, she completed a French-German master's degree in European Studies at the European University Viadrina and the Université de Strasbourg after a bachelor's degree in history at the University of Mannheim. Her interest in intercultural dialogue led to her current work with Dialogue with Science. The field of sustainability communication is new to her but very attractive as she is involved in an environmental organization in her spare time and tries to live as sustainably as possible.

Heike Janßen has worked as an editor since 1998 at ARD aktuell, the central newsroom of the national German public broadcasters. In 2019, she co-founded zwanzig50, a nonprofit for sustainability communication. She is also co-founder, board member, and since 2019 chair of the nonprofit association Netzwerk Weitblick – Verband Journalismus & Nachhaltigkeit, which trains media professionals in sustainability reporting and networks them with scientists. She also advises NGOs on public relations work in the field of development aid and sustainability. In addition, since 1990 she has worked as a cross-media freelance journalist in Africa and Eastern Europe. As a video journalist, she produces films for aid organizations and radio features for public broadcasters. Heike Janßen studied international politics in Hamburg and Windhoek and sustainability communication at Leuphana in Lüneburg. She learned storytelling in a 1-year course on screenwriting at the Film School Hamburg-Berlin.

Dorothee Menhart is a head of press and public relations at the nonprofit Wissenschaft im Dialog (WiD) in Berlin. This organization for science communication in Germany is active throughout Germany and supported by the major German research societies and foundations. It develops new communication formats that strengthen dialogue with society, with a focus on controversial topics and reaching new target groups. Throughout her work, the potential of storytelling has been a recurring theme. Before joining Wissenschaft im Dialog, she worked for 15 years in daily newspaper journalism after 2 years in advertising and an editorial traineeship, including as political editor and head of the university and science section of the newspaper Badische Zeitung in Freiburg. For a brief time, she also worked as an agency journalist. She likes science communication because there's something new every day. After all, almost nothing is as varied as science.

Petra Molthan-Hill, PhD, is a professor of sustainable management and education for sustainable development (ESD) at Nottingham Business School, Nottingham Trent University (NTU), UK, and co-chair of the United Nations Principles for Responsible Management Education (PRME) Working Group on Climate Change and Environment. She is an international multi-award-winning expert for Climate Change Mitigation Tools, Greenhouse Gas Management and ESD and leads the "Carbon Literacy Training for Educators, Communities, Organisations and Students" (CLT-ECOS) distributed worldwide. She is the editor of "The Business Student's Guide to Sustainable Management" and "Storytelling for Sustainability in Higher Education. An Educator's Handbook" both published by Routledge and led the Green Academy at NTU from 2013 to 2021.

Torsten Schäfer is a professor of journalism and text production at Darmstadt University of Applied Sciences, environmental journalist, and freelance writer. He teaches in the bachelor's program in Online Journalism, the master's program in Media, Technology, and Society, and heads the green journalism platform gruener-journalismus.de. His work is focused on environmental and climate journalism (research project Climate Stories from 2015 to 2019) as well as storytelling, stylistics, and EU reporting. He has received several awards and fellowships for journalism and research. From 2009 to 2013, Schäfer was editor of the international GEO edition and before that chief of staff and planner in the online editorial department of the international German broadcaster Deutsche Welle. Previous positions were political doctorate in EU reporting in German regional newspapers, master's degree in European Studies at RWTH Aachen University, and studies in journalism and political science at TU Dortmund University integrated with a newspaper traineeship. He was a freelance journalist for several years for the newspapers and news magazines GEO, taz, Süddeutsche Zeitung, and dpa.

Martin Storksdieck is the director of Oregon State University's STEM Research Center and a professor in OSU College of Education. The Center conducts applied research on STEM education and science engagement at the intersection of research, policy, and practice, with a strong focus on equity and social justice and the goal to align understanding of how all people learn throughout the lifespan in formal and informal settings with evidence-based strategies for STEM engagement. Storksdieck has more than 25 years of experience with research and evaluation in STEM-related fields and environmental and sustainability education. Before joining OSU, he directed the Board on Science Education and the Roundtable on Climate Change Education at the U.S. National Academy of Sciences. He currently serves on the Science Advisory Boards for the National Oceanic and Atmospheric Administration (NOAA) and the Leibniz Institute for Science and Mathematics Education in Kiel (Germany). He is also Chair of Trustees for TERC, a nonprofit R&D organization in Cambridge, MA, and serves as a board member of the Tree Media Foundation in Los Angeles, CA. He is an elected fellow of the American Association for the Advancement of Science. He holds master's degrees in biology and public policy, and a PhD degree in education.

Maren Urner holds a PhD in neuroscience and is a best-selling author, entrepreneur, and professor of media psychology. She studied and researched in Germany, Canada, the Netherlands, and the UK before co-founding *Perspective Daily* in 2016, the first German-language online magazine for constructive journalism. She is convinced that future-oriented reporting is needed, especially because of the climate catastrophe. To advance this goal in practice, she led *Perspective Daily* for 3 years as editor-in-chief and co-managing editor. Since October 2019, she has been a professor of media psychology at the HMKW University of Applied Sciences for Media, Communication, and Management in Cologne. Her 2019 bestseller with the title (in English) *Enough of the Daily Apocalypse* made her a sought-after interview guest for TV, radio, and podcasts. In March 2020, a German-language book she contributed to, *The Global Climate Emergency*, was published. In workshops and lectures, she advocates for brain-friendly and solution-oriented reporting that does not leave us helpless but motivates us to think and act sustainably.

Prologue

OF SUSTAINABILITY AND STORYTELLING – AN INTRODUCTION TO THIS BOOK

Anna Sundermann, Daniel Fischer, Sonja Fücker, and Hanna Selm[1]

Ekaterina[2] is a community development coordinator and would like to share experiences made by participants in a community climate project with others in her community. She wonders how she can best do that and persuade especially those who are skeptical about local climate protection initiatives of their value.

Stefan[2] promotes the One World Initiative and is an advocate for global justice. In a survey of his colleagues, he found that many thought their organizations were not as closely aligned with sustainability principles as they should be. He hopes that the storytelling approach will allow him to collect and share good examples with others in his network and creatively encourage people to do more to promote sustainability in their organizations.

Judith[2] is an instructor at a university. Last semester, she taught a seminar on the importance of forest gardens for biodiversity conservation. Together with her students, she now wants to share their findings with others. She wants to provide the students with material that will help them communicate their scientific findings in a way that people who have no prior knowledge of forest gardens or even biodiversity can understand the connections between biodiversity loss and sustainable development.

During our work at the Institute for Environmental and Sustainability Communication at the Leuphana University of Lüneburg in Germany, we have received these and numerous similar inquiries about how sustainability issues can be communicated in an interesting and, above all, effective way in different contexts. These inquiries reflect a growing interest in the storytelling approach by practitioners who have high expectations of the effects it will have and its effectiveness. These expectations were repeatedly connected with the belief that stories have an emotional impact and portray role models recipients can identify with, not only creating emotional closeness but also generating dismay, which would lead to changes in behavior.

DOI: 10.4324/9781003326144-2

The lively interest in and high expectations about its impact in practical fields of sustainability communication are dampened by a lack of systematic scientific studies that address the effects of sustainability-related storytelling. Scientific studies could provide us with practical, evidence-based recommendations for sustainability communication.

We want to contribute to narrowing this gap. The project *Storytelling in Sustainability Communication – Evidence and Perspectives* (short: Sustainability Storytelling, hereafter the SusTelling project), which was funded by the German Environmental Foundation between 2017 and 2021, is the basis of this book. Before we go into more detail about the project and the focus of this book, we want to take a step back to shed light on the challenging field of sustainability communication. This broader perspective will also highlight what efforts have already been made and where we would situate the storytelling approach in this field.

Challenges in Sustainability Communication

Even though several sustainability issues have emerged from their niches into mainstream awareness through political efforts and social movements over the past decade, the concept of sustainable development and its issues are still considered quite difficult to communicate (Djordjevic & Cotton 2011; Godemann 2021). Alone the question of what sustainability or sustainable development means has elicited, by the turn of the last century, more than 300 definitions, and no clear answer was in sight (Dobson 1996). What continues to characterize the challenges of sustainability communication is that we are confronted with complex, interrelated ecological and social systems in which individual actions are not directly linked to their potential impacts. Moreover, communication is complicated by the value-laden nature of the concept. For example, there are no clear answers to such questions as: What does justice mean for people today and in the future? Or: What are the basic human needs that must be met to live a good life? (Michelsen & Fischer 2017; Defila & Di Giulio 2020; Rauschmayer et al. 2020). Indeed, these questions must be discussed in society as a whole, since the consequences of any answer will affect society as a whole. It thus seems inevitable that the basis of persuasive and successful sustainability communication must be a societal discourse. Such a discourse is about more than communicating sustainability; it also requires a broad exchange about the rationales and values underlying the various understandings of sustainability (see Fischer et al. 2021).

Recently, the call for a society-wide discussion of sustainable development in countries on all continents has been amplified by several significant impulses. Many countries have developed national sustainable development strategies that are continuously updated, often accompanied by dialogue formats to engage the public (Meadowcroft 2007; Tils 2007). In September 2015, the 2030 Agenda for Sustainable Development was adopted by the United Nations, heralding a new phase in international sustainability policy (United Nations 2015). Comprising a total of 17 Sustainable Development Goals and 169 targets, it marked an expansion of the

Millennium Development Goals, which focused on reducing underdevelopment, and explicitly holds all countries accountable for contributing to global sustainable development. In the same year 2015, the Paris Agreement was adopted, which sets emission targets to be achieved through nationally determined contributions (Janetschek et al. 2020).

What these impulses have in common is that they see the transformation of society as a collaborative effort that requires to a large extent the activation and participation of civil society. It is a call for an informed civil society whose members have the needed knowledge, willingness, and skills to help shape the transformation toward a sustainable society (see, e.g., SDGs 4 and 12). Such mobilization calls for new "narratives of change" (Wittmayer et al. 2019) with creative forms of science communication that are easily understood and vivid. This approach could reduce fears of dynamic change and encourage people to overcome inertia and help shape a transformation toward a sustainable society (Lidskog et al. 2020).

Considering the challenges facing sustainability policymakers today, the main task is presenting the complexity and value of sustainability issues in such a manner that they are factually accurate and reliable – fundamental requirements for science communication – yet still reach their audiences. At the same time, the presentation of the issues should inspire and motivate them to participate in sustainable development. This is a demanding challenge. Media professionals, for example, complain that abstract complexity quickly places excessive demands on an audience and is met with incomprehension (Godemann 2021). Responses to this challenge include calls for replacing cognitively demanding communication formats with ones that are more unconventional and emotionalize the message (Reisch & Bietz 2007). Additionally, the formats should strike a balance between information and entertainment (Reinermann et al. 2014) and find their way (e.g., through pioneers of change) into everyday discourse through personalization (Jones & Peterson 2017). Communication must be seen as taking place among equals, presenting sustainability issues pragmatically and about everyday life, but without any "finger-wagging" (Reisch & Bietz 2007). The storytelling approach, considering these requirements, would seem to have considerable potential (Reinermann et al. 2014). This approach of using storytelling for sustainability communication, along with all the hopes associated with it, is the basis of the SusTelling project. Its results are detailed in the first part of this book and form a framework for the second part. We will now briefly explain the project's goals, structure, and approach, and then provide an overview of the contents of this volume.

The Storytelling in Sustainability Communication Project

The SusTelling project aims to place the storytelling approach on a systematic and evidence-based foundation and so make it fruitful for sustainability communication and able to contribute to a more effective presentation of sustainability topics in journalism as well as in the fields of science communication and corporate sustainability communication. The project focuses on communication with young people. The reason for this special focus is that young people play a key role in social change toward sustainable development. They are the ones, after all, who will have to face the future consequences of actions we take in the present. At the same time, young people are the ones who will be in decision-making positions in the future and who will have to deal with the many severe problems we leave behind. Sustainability communication with young people must consider their specific communication habits and needs. One particular challenge is to reach those in nonacademic milieus with a probably reserved attitude toward sustainability issues through new and activating forms of communication. The SusTelling project thus investigated storytelling as a way to encourage young people to engage with complex sustainability challenges such as climate change, species extinction, and social injustice.

The leading project question was broken down into three sub-questions that the research team at the Leuphana University of Lüneburg then tackled during the 3-year project period, with the support of three experts from the fields of psychology, narratology, and journalism as well as a project advisory board. The three work phases are briefly outlined in the following.

What Is Our Understanding of Sustainability-Related Storytelling?

The entire duration of the project was filled with discussions on the fundamental question of how to analytically define storytelling and what special requirements are connected with its use in sustainability communication. Valuable impulses for further developing our conceptual understanding came from presentations at professional conferences, a systematic literature analysis on the effects of storytelling, and the empirical studies we conducted. These impulses allowed our understanding of SusTelling[3] to continue to develop, which can be seen in the different emphases in project publications (see Fischer et al. 2020 and Chapter 1).

What Effects Does Sustainability-Related Storytelling Have?

Empirical investigations of the effects of sustainability-related storytelling were at the center of the project. As a first step, the project included a systematic literature review of the reported effects of storytelling in different scientific fields such as health communication, marketing, and educational research. Preparing for our research, six expert interviews were conducted with journalists to identify assumptions associated with the impact of storytelling in the field. The results of the two studies were then compared and formed the empirical basis for the project.

After sharpening our theoretical concepts, our next step was to review and analyze existing narrative media messages from the field of sustainable consumption. We slightly redesigned and modified individual messages to use them in the experimental study (Sundermann et al. 2019). The remaining messages were prepared for use in the exploratory study. Both studies used different approaches to examine the effects of SusTelling-style media messages. In doing so, they focused on academic and nonacademic educational backgrounds in various disciplines and professions and on people with lower or higher levels of interest in sustainability issues. They are described in detail in Chapters 2 and 3 of this volume.

How Can Sustainability-Related Storytelling Be Taught, Learned, and Practiced?

In the SusTelling project, we did want not only to produce new knowledge but also to make it usable in various fields of practice. Transfer activities were therefore an integral and comprehensive part of the project. Our strategy was developed with the participation of various practitioners and academic researchers. This included a biannual newsletter, Twitter activities, a virtual closing event, presentations at conferences and expert forums, the development of an online learning unit, and this volume. Reflections on transferring SusTelling into practice are discussed in the form of a conversation in Chapter 4.

About This Volume

This volume is composed of two parts. In the first part (see Chapters 1 to 4), we discuss the project results. In the second part of the volume (see Chapters 5 to 8), we give examples of how storytelling is used in sustainability communication from the perspectives of different fields of practice. All the chapters in this volume respond to one of the three project questions.

What Is Our Understanding of Sustainability-Related Storytelling?

The definition of the term and our understanding of storytelling for sustainability are mainly found in Chapter 1. This understanding is one of the results of our project and forms the basis for the empirical findings presented in Chapters 2 and 3 as well as for the development of an online self-learning unit, which is explained as a transfer product in Chapter 4. The chapters in Part 2 of the volume each draw on the understanding of SusTelling presented in Chapter 1 and place it against a background in a field of practice.

What Are the Effects of Sustainability-Related Storytelling?

Results on the effectiveness of sustainability-related storytelling can be found in Chapters 2 and 3. Each of these two chapters describes the findings of a study and shows how it can be used in communication practice. For readers, these two chapters offer insights into whether and how storytelling can work in different target groups and how it invites reflection and engagement in different ways. Specific references to scientific findings on the impact of storytelling are also made in the other chapters, for example, when results of cognition research are summarized to inform constructive journalism (see Chapter 6) or findings of teaching–learning research in higher education (see Chapter 5).

How Can Sustainability-Related Storytelling Be Taught, Learned, and Practiced?

Answers to questions about the implementation of SusTelling in education, training, and professional practice are mainly found in Chapters 4–8. Chapter 4 presents two approaches to teaching storytelling for sustainability and discusses challenges and recommendations in the form of a scripted conversation. In Chapters 5–8, practitioners have their say, taking a practical perspective from their field. Using concrete examples, these chapters show how storytelling is and should be used in journalism, science communication, university teaching, and corporate communication, without losing a critical focus on the potential risks and side effects of narrative forms of sustainability communication.

This volume represents an attempt to describe the effects and uses of storytelling in sustainability communication and to make the resulting insight, including those that are critical, fruitful for different fields of practice. In this way, we hope to contribute to the reflective use of storytelling in communication on sustainability issues, one based not on the effects it is assumed to have but on scientific findings and well-founded practical experience. The diverse examples and practical experiences collected in this volume indicate a wealth of possible uses – but also show how great the need is for further research into the use of storytelling for sustainability.

Acknowledgments

Our work on the SusTelling project and this volume would not have been possible without the support of many dedicated people.

First, we would like to thank the German Environmental Foundation and Dr. Thomas Pyhel for their funding and their constant constructive and critical support of our project. Over the course of several years, we received important impulses from the members of the project advisory board. We would like to thank Thomas Friemel, Prof. Dr. Maximilian Gege, Maike Gosch, Prof. Dr. Sarah Jastram, Dorothee Menhart, Prof. Dr. Gerd Michelsen, Dr. Thomas Pyhel, and Prof. Dr. Maren Urner for so generously giving us their expertise and time. The implementation of a 3-year project with research, practice, and transfer parts across different countries and time zones can only succeed in teamwork. We would like to thank our experts Prof. Dr. Daniel Hanss, Dr. Elisabeth Hollerweger, and Prof. Dr. Torsten Schäfer for their valuable input. Throughout the project, we were supported at various points by student assistants. In the first half of the project by Lukas Heck, Tanja Montag, and Henriette Reinhardt, and later mainly by Sonja Fischer, Anna Schmidt, Judith Sievers, and Julia Silver. We would also like to thank them for their efforts in making the project a success. Special thanks are due to the young people who made our research possible in the first place by participating in our investigations, as well as to the many media professionals who made their SusTelling stories available to us free of charge. Furthermore, we are indebted to our numerous network partners who promoted the transfer of our project, as well as to Denisse López, Lars Fabian Paape, Kai-Daniel Du, and Katinka Malena Mustelin for their professional support in implementing the SusTelling online learning unit.

Finally, this volume would not be possible without the contributing authors. Many sincere thanks for their contributions. We would also like to thank the publishing houses oekom and Routledge for their professional support, especially Annabelle Harris and Jyotsna Gurung for their guidance in getting this international volume published. Furthermore, we would like to thank Dr. Elisabeth Hollerweger, whose thorough and critical (in the best sense of the word) editing improved this volume. We want to express our gratitude to Paul Lauer for the translation of the manuscript, Markus Miller for the translation of the artwork, and Sophie van Gent for the editorial support. Needless to say, if any errors should have crept in, these have to be attributed entirely to the editors.

Notes

1 The authors contributed equally to the prologue and epilogue and to the curation of the volume.
2 Name changed by the authors.
3 The terms "SusTelling", "sustainability-related storytelling", and "storytelling for sustainability" are used interchangeably in this book. Whenever the research project is mentioned, this is always indicated by adding the term "SusTelling project".

References

Defila, R., & Di Giulio, A. (2020). The concept of "consumption corridors" meets society: How an idea for fundamental changes in consumption is received. *Journal of Consumer Policy*, *43*(2), pp. 315–344.

Djordjevic, A., & Cotton, D. (2011). Communicating the sustainability message in higher education institutions. *International Journal of Sustainability in Higher Education*, *12*(4), pp. 381–394.

Dobson, A. (1996). Environment sustainabilities: An analysis and a typology. *Environmental Politics*, *5*(3), pp. 401–428.

Fischer, D., Reinermann, J.-L., Guillen Mandujano, G., DesRoches, C. T., Diddi, S., & Vergragt, P.J. (2021). Sustainable consumption communication: A review of an emerging field of research. *Journal of Cleaner Production*, *300*, 126880.

Fischer, D., Selm, H., Sundermann, A., & Storksdieck, M. (2020). Storytelling for sustainability: A definition and its implications for education and communication. In P. Molthan-Hill, D. Baden, T. Wall, H. Puntha, & H. Luna (eds.), *Storytelling for Sustainability in Higher Education: An Educator's Handbook* (pp. 38–51). London: Routledge.

Godemann, J. (2021). Communicating sustainability. Some thoughts and recommendations for enhancing sustainability communication. In F. Weder, L. Krainer, & M. Karmasin (eds.), *The Sustainability Communication Reader* (pp. 15–29). Wiesbaden: Springer.

Janetschek, H., Brandi, C., Dzebo, A., & Hackmann, B. (2020). The 2030 Agenda and the Paris Agreement: Voluntary contributions towards thematic policy coherence. *Climate Policy*, *20*(4), pp. 430–442.

Jones, M.D., & Peterson, H. (2017). Narrative persuasion and storytelling as climate communication strategies. In *Oxford Research Encyclopedias: Climate Science*. Oxford: Oxford University Press.

Lidskog, R., Berg, M., Gustafsson, K.M., & Löfmarck, E. (2020). Cold science meets hot weather: Environmental threats, emotional messages and scientific storytelling. *Media and Communication*, *8*(1), pp. 118–128.

Meadowcroft, J. (2007). National sustainable development strategies: Features, challenges and reflexivity. *European Environment*, *17*(3), pp. 152–163.

Michelsen, G., & Fischer, D. (2017). Sustainability and education. In M. v. Hauff & C. Kuhnke (eds.), *Sustainable Development Policy: A European Perspective* (pp. 135–158). London: Routledge.

Rauschmayer, F., Masson, T., Leßmann, O., & Gutwald, R. (2020). Sustaining human well-being across time and space – sustainable development, justice and the capability approach. In A. Crabtree (ed.), *Sustainability, Capabilities and Human Security* (pp. 75–102). London: Palgrave Macmillan.

Reinermann, J.-L., Lubjuhn, S., Bouman, M., & Singhal, A. (2014). Entertainment-education: Storytelling for the greater, greener good. *International Journal of Sustainable Development*, *17*(2), pp. 176–191.

Reisch, L.A., & Bietz, S. (2007). How to convince the unconvincibles? A mass mediated approach to communicate sustainable lifestyles to a low-interest target group. *International Journal of Innovation and Sustainable Development*, *2*(2), pp. 192–200.

Sundermann, A., Selm, H., Fischer, D., Silver, J., & Fischer, S. (2019). *Technical Report on the Experimental Study of the SusTelling Project*. Lüneburg: Leuphana Universität Lüneburg.

Tils, R. (2007). The German sustainable development strategy: Facing policy, management and political strategy assessments. *European Environment*, *17*(3), pp. 164–176.

United Nations (2015). *Transforming Our World: The 2030 Agenda for Sustainable Development: Resolution Adopted by the General Assembly on 25 September 2015*. New York: United Nations.

Wittmayer, J.M., Backhaus, J., Avelino, F., Pel, B., Strasser, T., Kunze, I., & Zuijderwijk, L. (2019). Narratives of change: How social innovation initiatives construct societal transformation. *Futures*, *112*, 102433.

PART I
Research Perspectives

1

SUSTELLING

Storytelling for Sustainability

Daniel Fischer, Sonja Fücker, Hanna Selm, Martin Storksdieck, and Anna Sundermann

Let us begin this chapter with a question: What do the following three examples have in common?

1. The packaging of some processed Danish meat states that this product contains ham naturally preserved with Nordic herbs, as has been practiced in the North for centuries according to folk stories (Fenger et al. 2015).
2. A social worker in the southwestern United States goes into Latino neighborhoods and uses the personal story of a family to describe why regular screening is important for colorectal cancer prevention (Larkey & Gonzalez 2007).
3. Employees are asked to watch a video that shows vignettes of a fictitious manager and afterward evaluate this manager's authenticity. The manager's perceived authenticity increased the more they explained their management style by making connections to their life history, in particular, to negative turning points in their biography (Weischer et al. 2013).

All three examples come from academic publications investigating the effect of storytelling. The three examples from product marketing, health promotion, and management research already hint at what is characteristic of storytelling as a field of research and practice – even though there are extremely diverse views on how we may understand and use storytelling. In this chapter, we are connecting storytelling with sustainability, to see what specific requirements must be met by storytelling for sustainability.

4. As part of the *Model London* game, the *Stories for Change* project invited visitors to a museum to develop short utopian stories about how London's energy problems could be solved in the future. The stories were modeled with Lego

DOI: 10.4324/9781003326144-4

figures and rated by game participants to identify the "least bad" ideas (Smith et al. 2017).

As the fourth example shows, storytelling about a sustainability-related topic (in this case, the future of energy supply) pursues specific goals. It is used to develop very specific approaches to complex and knotty (wicked) sustainability challenges, and it is used to engage people with these complex problems by involving them in producing possible solutions. We will discuss this requirement later. First, we will discuss a broad range of perspectives, conceptual explanations, and definitions in the current research literature to derive a more comprehensive understanding of storytelling (1.1), to subsequently focus this understanding on the field of sustainability communication (1.2), and then discuss possible uses of storytelling for sustainability (1.3).

1.1 What Is Storytelling?

Storytelling is about using narrative structures to achieve certain affective, cognitive, or conative effects in readers. To sharpen the meaning of the term, the following section distinguishes between the characteristics of stories (their "inner" characteristics) and the contexts in which these stories are used and told (their "outer" characteristics).

1.1.1 Inner Characteristics of a Story

We understand storytelling as a communicative practice in which content (facts and information) is transformed into a narrative form that we call "story". Stories have a temporal order (*chronology*) and usually a recognizable beginning, middle, and end (Nwoga 2000; Hinyard & Kreuter 2007). This temporal order does not always have to be linear. For example, "time jumps" such as flashbacks or flash-forwards are common structuring elements of narrative time. Stories can also be told without a linear temporal structure, for example, when different events unfold one after the other and the actual chronological course of events is not immediately revealed. Stories also provide information about the temporal and spatial environment in which they take place (Woodside 2010). But perhaps the most central and obvious characteristic of a story is its *plot*, which typically revolves around central characters (*personalization*), who may be humans or other living creatures (real or fantasy), or even technological devices (think of the robot in Disney's science fiction animated film *Wall-E*). The plot develops around the characters' experiences and conflicts, often following archetypal patterns of how a plot unfolds in narrative structures.

Various typologies of the most common plotlines have been developed and proposed in the literature (Tobias 2011), based on an underlying theme, such as adventure, discovery, escape, love, transformation, redemption, or rise and fall. While individual plots can also be combined, each of them provides a basic framework for the narrative; each plot contains its own specific chain of events, character configurations, and space-time structure (*dramaturgy*). Other characteristics of stories include media-specific representational elements (*stylistics*) such as linguistic, pictorial, or musical devices that help, for example, to create suspense or evoke emotions (Früh 2014; Emde et al. 2016). Stylistic – sometimes called literary or rhetorical – devices such as alliteration and metaphor have been extensively studied and described in literary studies (Wales 2011). Depending on the goal underlying their use, a narrative can also take on different tones that convey mood and create an atmosphere (*tonality*). In entertainment-oriented storytelling, for example, the tonality can be humorous and amusing or dramatic and frightening, whereas, in informative or so-called serious storytelling, a more educational or content-oriented approach is used to activate learning processes (Downs 2014). The tone of storytelling can vary depending on whether the goal is to persuade or deliberate. Persuading an audience to adopt a point of view or attitude is accomplished through stylistic and rhetorical means. This is especially true of storytelling containing moral messages (Lee & Leets 2002), for example, Al Gore's *An Inconvenient Truth* or David Attenborough's documentary *A Life on Our Planet*, in which he shares his concern for the current state of the planet. In contrast, approaches that seek to promote processes of understanding tend to use a more balanced and conversational tonality (Arnett et al. 2008) by raising questions or presenting competing points of view, for example. Depending on the plot, the tone can be more negative (problem-focused) or positive (constructive, solution-oriented) (Lengauer et al. 2012).

The tone in stories is often associated with different ways of engaging the audience in the story (*modality*). In interactive storytelling, for example, the audience has the opportunity to interact with the media content and direct the discussion itself (Baldwin & Ching 2017). Examples include interactive exhibitions or online social media. The social media storytelling approach focuses on personal interactivity. In a project called *Time Capsules*, a series of provocative and one-dimensional statements on social or environmental issues was disseminated through social media using a fake sender to trigger reactions and stimulate interaction processes (Vervoort et al. 2014). The worldview expressed in these postings challenged the audience to clarify their own viewpoints and engage with counter-narratives (so-called scenario storytelling). Storytelling can also provide highly immersive experiences to recipients. Such innovative forms of storytelling have proliferated in recent years with the availability of augmented reality and virtual reality technologies. For example, a research team at Stanford University used short (8–10 minutes) virtual reality scenarios at the *Tech Museum of Innovation* in San Jose, California, to promote empathy with homeless people (Herrera et al. 2018). A recent European example is the virtual reality exhibition about the Aletsch Glacier (expedition2grad.ch).

1.1.2 External Characteristics of Storytelling

In addition to these internal characteristics of stories, it is helpful to be aware of the contexts in which storytelling is embedded – not least because these contexts can significantly affect the way the internal characteristics play out (van Hulst & Vbema

2020). These contextual layers are interconnected or nested so that they form an external environment in which the inner story can unfold. We distinguish between three such contextual layers: the setting, the medium, and the topic.

The first distinguishing characteristic of storytelling approaches is the place, scene, or social context in which storytelling is used to communicate with an audience (*setting*). Settings in which storytelling approaches are commonly used include journalism and science communication, formal and informal education (including museums and exhibitions), public relations (PR), and health communication, as well as change management and corporate communication. The intentions behind storytelling in these different settings vary widely, as the examples presented at the beginning of this chapter show, but they are focused on knowledge acquisition or behavioral change.

The second distinguishing feature relates to the different means of communication used in storytelling (*medium*). Oral storytelling is the oldest and most natural form of communicating information. Media as a storytelling context describes the communication channel (or combination of different channels) used to reach an audience. Communication channels are "the means by which messages travel from one individual to another" (Rogers 2003). Today, storytellers use a variety of media, from oral speech to videos (YouTube, TikTok, and the like increasingly complement and replace traditional film formats), to social media such as Instagram, Facebook, and LinkedIn. Short messaging services such as Twitter are also often used to attract attention and gain audiences. Stories are not only conceived for or transformed into different media; they are sometimes designed to unfold as a coherent narrative only through the interaction of different media. This so-called cross-media or transmedia storytelling today mostly involves digital storytelling (Giovagnoli et al. 2014). One example is the Brazilian *telenovelas* about public health between 1975 and 1982, which were a commercial success and led to greater use of family planning methods and progress in gender equality (Singhal 2013). Different narrative elements were creatively coordinated across different communication platforms (transmedia) to build a story world and increase the entertainment value for audiences (Jenkins 2007). Such a coherent and integrated form of transmedia communication should be distinguished from the use of different media formats to convey the same content across different media, for example, when a newspaper article is also published in an online version that is easily shared and commented on by readers using social media.

Finally, storytelling is often used to take up specific content (phenomena or objects) in a particular way and communicate it to an audience (*theme*). It is important to distinguish between content and theme (Haan 2002): a content (ingredients of a food product or a medical service, as in our examples at the beginning of this chapter) can be framed or illuminated from different perspectives (in our examples, as a regional and culturally traditional product or as a health-preserving measure). Concerning sustainability, a large number of possible important themes can be identified if one thinks of the fields of action addressed in the Sustainable

Development Goals. There are, however, many other themes that can be used to frame a specific content as a sustainability topic.[1]

1.2 What Is SusTelling?

Sustainability-related storytelling, in the understanding we propose, is a type of sustainability communication that educates and promotes change toward sustainable development. On the basis of these objectives (1.2.1) and the internal and external characteristics we have just elaborated on, we put forward a definition of storytelling for sustainability, SusTelling for short (1.2.2), and illustrate it with a case study (1.2.3).

1.2.1 Normativity as a Defining Characteristic of SusTelling

While the characteristics of storytelling outlined earlier seem to represent a certain interdisciplinary consensus, there is no agreement on what makes storytelling in the context of sustainability communication special or different. However, connecting the general characteristics of storytelling to the scholarly discourse of the last 25 years on sustainability communication and education can serve as a starting point. When storytelling is used to address sustainability in an educational and communicative context, it almost inevitably raises questions about normativity.

Sustainability itself is a normative idea as it refers to fundamental principles of inter- and intragenerational justice and the "possibility that human and other life will thrive on the planet forever" (Ehrenfeld 2008). Storytelling in sustainability communication should therefore produce effects that are conducive to sustainable development (e.g., audience intentions to consume more sustainably in the future). Sustainability communication, in its theoretical orientation, aims to provide knowledge about human–nature interactions and to strengthen awareness of social problems (Godemann & Michelsen 2011). It should help people to better adapt to current and upcoming challenges and enable them to actively participate in shaping a sustainable society. In this understanding informed by educational theory, sustainability communication builds on the principles of reflexivity and emancipation. Reflexivity is the ability to take detached perspectives, reflect on oneself, and explore individual and collective development processes. Emancipation refers to the ability to act freely, distance oneself from negative developments in life, and experience autonomy over one's own trajectory (Masschelein 2004). To be educational, storytelling should foster these reflexive and emancipatory capacities and not stop at "nudging" people to adopt certain opinions or behaviors.

There are then two normative demands placed on storytelling in the context of sustainability. It should achieve very concrete sustainability effects (improvements in environmental and social conditions), and it should empower people to act in a self-determined and competent manner for the benefit of sustainable development.

1.2.2 SusTelling: A Definition

Our definition of SusTelling brings together the essential characteristics of storytelling described earlier with the double normativity inherent in the concept of sustainability and the theoretical foundations of sustainability communication. Ideally, SusTelling has the following eight characteristics which serve to foster two main outcomes, to be educationally formative for the recipient (*educational outcome*) and to promote sustainability (*sustainability outcome*):

1. a clear arrangement of action sequences (*plot*)
2. a focus on characters and their experiences (*personalization*)
3. conflicts, developments, and solutions (*dramaturgy*)
4. a specific temporal arrangement (*chronology*)
5. a temporal, spatial, or cultural environment (*context*)
6. an arc of tension created through stylistic devices (*stylistics*)
7. a certain mood (*tonality*)
8. a more or less interactive or immersive way of presentation (*modality*)

In summary, we view SusTelling as a multifaceted practice that reflects the internal and external characteristics mentioned earlier, as well as the dual goals of educating audiences and promoting change toward sustainable development (Figure 1.1). SusTelling addresses core questions and research fields in sustainability

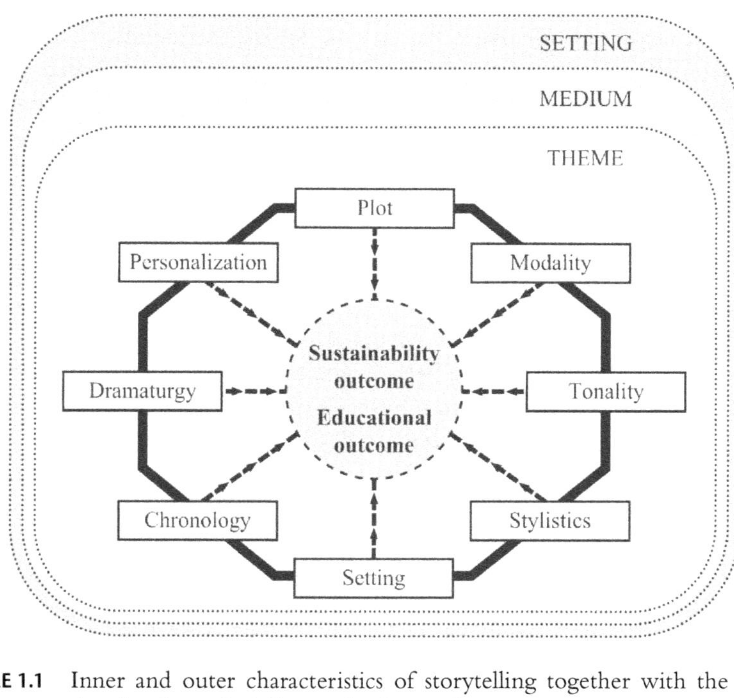

FIGURE 1.1 Inner and outer characteristics of storytelling together with the educational and sustainability goals of SusTelling

science, as discussed in the literature (see, e.g., Kates 2011). Individual behavior includes various forms of social and political engagement or everyday consumption practices that impact sustainability, for example, in the areas of nutrition, mobility, or housing (Geiger et al. 2018; Micheletti & Stolle 2012). In this context, the idea of sustainability differs from many other topics due to its complexity and ambiguity, as often research findings can offer no clear guidelines for action (e.g., are organically grown vegetables from faraway countries more sustainable than conventionally grown ones from the region?).

1.2.3 A SusTelling Case Study: There's a Rang-Tan in My Bedroom

We want to explain our understanding of SusTelling with a case study. We suggest you first watch the one-and-a-half-minute video clip *There's A Rang-tan in My Bedroom* (Alpin 2018), narrated by Emma Thompson, and then proceed to read later (https://bit.ly/3t0JT9C).

The video was produced for a Greenpeace UK campaign for use on various channels (*medium*) to raise public awareness of the consequences of palm oil deforestation. However, it was also used by a drugstore chain to draw attention to the importance of avoiding palm oil products (*context*). Concerning content, the video deals with the destruction of the orangutan's habitat for the cultivation of palm oil by humans (*theme*).

The structure resembles that of a classic fable (*plot*). The starting point of the story is a conflict between a girl and an orangutan, which is emblematic of the violent intrusion of humans into the habitat of orangutans. This negative experience with the orangutan in her bedroom gives the girl, who is at first frightened and angry, the possibility to adopt the animal's perspective and participate in its reality. The audience experiences how an intrusion into their private sphere is an unpleasant experience (*personalization*). In the course of the story, the girl learns why the orangutan was acting so inconsiderately. The video ends with the girl solving the conflict by starting a petition among her friends to stop the destruction of the orangutans' habitat as the rainforest is being destroyed to cultivate palm oil (*dramaturgy*).

Regarding temporal structure, there is a clear beginning in the narrative: the girl is annoyed by an orangutan that has made her bedroom its new home and is destroying her things. In the middle of the story, a flashback to the orangutan's natural habitat shows the girl how it is being destroyed by humans. This shows the consequences of palm oil cultivation. At the end of the story, the girl decides to help save the orangutan's habitat by fighting against the palm oil industry and motivates others to join her (*chronology*). The video gives clear information about the settings of the story and differentiates them through coloring and variations in music (*setting*). Some stylistic devices are used in the video to reinforce the message. Allegory plays with the role reversal of orangutan (animal) and girl (human), making the problem accessible, plausible, and understandable. Moreover, the cartoon technique at the beginning is reminiscent of a children's movie, reflecting the girl's initial childlike point of view. Sentence beginnings are also repeated, and the dialogue between the girl and the orangutan is written in rhyme. This makes the narrative particularly memorable (*stylistics*).

Rhymes are often associated with a lively and cheerful mood, such as the one right at the beginning of the video. However, the music changes to emphasize the narrative's dramaturgy. The destruction of the rainforest is represented by a change to a threatening and somber mood, emphasized by the dark colors and the shift to black-and-white presentation (*tonality*). The video calls for the viewer to take part in a Greenpeace campaign and thus aims to have a motivating effect on the audience, but beyond that, it does not have any interactive elements, such as the possibility of entering into a dialogue with the girl or changing the plot (*modality*). The message at the end of the story calls for collective action against rainforest deforestation. Activism as the solution is underscored by the fact that the girl is portrayed in a brighter light in the scene, where she appears as a change agent and a role model for other children. The video thus conveys a clear message about protecting the environment and wildlife through social action. It draws attention to the conflictual relationship between the palm oil industry, the decimation of the orangutan population, and the consumption of chocolate and shampoo (*normativity*).

1.3 How Can SusTelling Be Used?

From the literature on storytelling, previous work on sustainability communication, and the few works that discuss storytelling in the context of sustainability, we have a good idea of what the potential and opportunities, but also the dangers and risks, are with using storytelling in sustainability communication.

1.3.1 Opportunities

Narrative approaches such as storytelling open up a range of new opportunities for sustainability communication. Four stand out. First, through its narrative patterns and forms, storytelling provides a framework that draws on long-practiced human traditions of knowledge transmission and thus follows well-known patterns of reception. Stories serve as archetypal frameworks for humans to interpret and assimilate complex thoughts (Sugiyama 1996). This fundamental connection between our perceptual capacity and stories offers the opportunity to capture attention, reduce the cognitive effort required for reception, and provide the means to overcome cultural barriers – for example, when communicating with educationally disadvantaged people who may be initially less receptive to scientific knowledge and issues (see, e.g., Rosin et al. 2019).

Second, narrative forms provide possibilities to shed light on complex subject matters. They allow abstract concepts such as, for example, the life cycle of consumer goods to be presented in a way that allows the interrelationship of consumption and production systems to unfold organically as part of the plot. This was done in *The T-shirt Under the Southern Cross*, a 60-minute story staged in a planetarium in the late 1990s. The story is about the role of astronomy in enabling global trade with unforeseen environmental consequences (such as modern T-shirts traveling back and forth between continents as part of the production process).

Third, stories can evoke moral judgments, opening up a normative resonance space by confronting the characters in the story with difficult choices based on conflicting value systems, such as in the film *Captain Fantastic* (released in 2016). The film is about a couple devoting themselves to raising their six children off the grid until the father is forced by his wife's death of cancer to enter the modern world of consumerism. The modern world challenges his notion of what it means to act sustainably and prepare his children for their future. Storytelling provides an ideal platform to explore the value dimension of sustainability and the normative foundations of sustainability itself (such as intra- and intergenerational justice).

Fourth, storytelling offers the potential to explore human experiences of everyday life, to help audiences understand and empathize with others' life experiences so that American or European audiences, for example, can share in the lived experiences of other peoples. Narrative forms can therefore help individuals explore and find new opportunities and entry points for engaging with the complexity of sustainability in their personal or political lives. Storytelling thus allows for an empathic sharing of emotions and a deep personal connection to sustainability issues, as illustrated by the examples of the Greenpeace video clip *There's A Rang-tan in My Bedroom* or the full-length film *Captain Fantastic.*

In summary, stories provide a way for people to engage with complex contexts and moral–ethical issues and, in doing so, learn about new ways of looking at things. An example of how stories can address issues of sustainable development is Heinrich Böll's *Anecdote Concerning the Lowering of Productivity*. The story uses the form of a parable, ostensibly to depict a conversation between a tourist and a fisherman about the benefits of industrial fishing. The seemingly mundane event recounted in the story is essentially a fundamental critique of Western thinking about the complex relationship between economic growth, productivity, and quality of life. It is therefore not surprising that it premiered on radio in 1963 on Labor Day of all days (Böll 1995) and, given how it raises fundamental questions in an accessible way, it still has relevance today.

1.3.2 Risks

However, using storytelling as a tool also poses some risks for sustainability communication. Three risks seem to be particularly important: misunderstandings due to an excessive reduction of complexity, conflicting goals by conveying unintended messages, and its functionalization for purposes that run counter to the normative goals of sustainability communication, that is, motivating and empowering the audience to act.

Evidence of unintended and undesired consequences are found in sustainability assessment and social marketing, where a necessary reduction of complexity can lead to reduced accuracy and oversimplification (Bond & Morrison-Saunders 2011), which in turn lead to misunderstandings about the foundations of sustainability itself (Choi et al. 2010). For example, ignoring the dynamics and interactions of different factors affecting systems in favor of isolated causal mechanisms makes it easier to focus on specific environmentally friendly behaviors. (Bring your own shopping bags to avoid polluting the ocean with plastic.) Such simplistic messages may fail to address underlying, systemic causes (Why is plastic everywhere in our lives?). This pitfall can best be avoided by linking linear arguments (Bring your own grocery bags, so you don't need plastic or paper bags) back to a systems-level discussion of sustainability (What role do plastic bags really serve? What are the alternatives? What are the consequences are of continued reliance on plastic bags and plastic packaging?).

Problematic consequences in storytelling can also arise from the use of archetypal narrative forms. McGreavy and Lindenfeld (2014) demonstrate how climate change documentaries while having a positive impact on audience awareness of climate change issues can have unintended consequences. For example, the study shows that the choice of a particular storytelling frame (comedy with exaggerated gender stereotypes) had negative effects on other sustainability goals (in this case, gender equality). Thus, contradictory messages are conveyed in a subtle way.

The third risk of using archetypal narrative forms is to disempower audiences by reducing complexity to simplistic truths or plots. The potential of storytelling can be diminished by conveying a particular concept or changing a particular behavior in a particular audience (e.g., the typical encouragement to recycle). However, such an instrumental "transmission logic" contradicts the normative understanding of sustainability communication. People should be empowered to participate in society as critical actors able to make decisions on the basis of evidence and their values. Campaign-like instrumental approaches to storytelling designed to achieve clear predetermined change goals run the risk of steering people's behavior rather than empowering them to make competent choices. This is not to doubt the legitimacy of all social marketing campaigns promoting an environmentally friendly action; it is just that such a narrow understanding of storytelling conflicts with the objectives of sustainability communication as outlined here. We should be critical of storytelling when it begins to obscure the complexity of issues and distract audiences from independent thought in favor of manipulating behavior change and providing simplistic instructions for action.

1.4 Conclusion

Although storytelling itself has a long-standing tradition and there is now experience and knowledge in various fields of research and practice about how storytelling works and what it can achieve, SusTelling – or storytelling for sustainability – represents a very young field. In this chapter, we have argued that SusTelling is a concept that reflects general internal and external characteristics of storytelling and specifies the normativity implicit in the idea of sustainability. We believe that the understanding we propose here can help to explore storytelling in a sustainability context more systematically and to practice it more strategically. This requires not only knowledge about conceptualization but also empirical evidence of how different forms of SusTelling affect recipients. Although storytelling holds promising potential for sustainability, there is currently a lack of solid evidence on the impact of this approach on audiences. In particular, a differentiated view of SusTelling effects is needed, for example, regarding the sociodemographic or sociocultural backgrounds of recipients, the quality, and quantity of media exposure, and the temporal dimension of effects (i.e., how stable they are). These research gaps represent a major obstacle to a more comprehensive, competent, and informed use of storytelling in sustainability communication. The following chapters on empirical research from the SusTelling project contribute to filling these gaps and promoting an evidence-based professionalization of storytelling in sustainability communication.

Editorial Note on the Text

This chapter synthesizes and updates contributions published in other places to conceptualize an understanding of storytelling in sustainability communication (Fischer & Storksdieck 2018; Fischer et al. 2020). We would also like to thank Carolin von Kameke and Robin Marwege for their contributions to the groundwork of the research project, on which this chapter is also based.

Note

1 In what follows, we will also use the term "topic" to refer to content that has been addressed under a sustainability theme.

References

Alpin, S. (2018). *Rang-tan Greenpeace.* [https://bit.ly/3wUhxjN]. Accessed on May 5, 2022.

Arnett, R. C., Arneson, P., & Holba, A. (2008). Bridges not walls: The communicative enactment of dialogic storytelling. *Review of Communication, 8*(3), pp. 217–234.

Baldwin, S., & Ching, Y.-H. (2017). Interactive storytelling: Opportunities for online course design. *TechTrends, 61*(2), pp. 179–186.

Böll, H. (1995). Anecdote concerning the lowering of productivity. In L. Vennewitz (ed.), *European Classics. The Stories of Heinrich Böll* (pp. 628–630). Evanston: Northwestern University Press.

Bond, A. J., & Morrison-Saunders, A. (2011). Re-evaluating sustainability assessment: Aligning the vision and the practice. *Environmental Impact Assessment Review, 31*(1), pp. 1–7.

Choi, S., Niyogi, D., Shepardson, D. P., & Charusombat, U. (2010). Do earth and environmental science textbooks promote middle and high school students' conceptual development about climate change? *Bulletin of the American Meteorological Society, 91*(7), pp. 889–898.

Downs, J. S. (2014). Prescriptive scientific narratives for communicating usable science. *Proceedings of the National Academy of Sciences of the United States of America, 111*(Suppl. 4), pp. 13627–13633.

Ehrenfeld, J. (2008). *Sustainability by Design: A Subversive Strategy for Transforming Our Consumer Culture.* New Haven: Yale University Press.

Emde, K., Klimmt, C., & Schluetz, D. M. (2016). Does storytelling help adolescents to process the news? *Journalism Studies*, *17*(5), pp. 608–627.

Fenger, M. H. J., Aschemann-Witzel, J., Hansen, F., & Grunert, K. G. (2015). Delicious words – Assessing the impact of short storytelling messages on consumer preferences for variations of a new processed meat product. *Food Quality and Preference*, *41*, pp. 237–242.

Fischer, D., Selm, H., Sundermann, A., & Storksdieck, M. (2020). Storytelling for sustainability: A definition and its implications for education and communication. In P. Molthan-Hill, D. Baden, T. Wall, H. Puntha, & H. Luna (eds.), *Storytelling for Sustainability in Higher Education* (pp. 38–51). London: Routledge.

Fischer, D., & Storksdieck, M. (2018). Storytelling: Ein Ansatz zum Umgang mit Komplexität in der Nachhaltigkeitskommunikation [Storytelling: An approach to deal with complexity in sustainability communication]. In T. Pyhel (ed.), *Zwischen Ohnmacht und Zuversicht? Vom Umgang mit Komplexität in der Nachhaltigkeitskommunikation* [Between Paralysis and Optimism? Dealing With Complexity in Sustainability Communication] (pp. 161–176). Munich: Oekom.

Früh, W. (2014). Narration und storytelling [Narration and storytelling]. In W. Früh & F. Frey (eds.), *Narration und Storytelling. Theorie und empirische Befunde* [Narration and Storytelling. Theory and Empirical Evidence] (pp. 63–119). Cologne: Herbert von Halem Verlag.

Geiger, S. M., Fischer, D., & Schrader, U. (2018). Measuring what matters in sustainable consumption: An integrative framework for the selection of relevant behaviors. *Sustainable Development*, *26*(1), pp. 18–33.

Giovagnoli, M., Vaglioni, P., & Montesano, F. (2014). *Transmedia Storytelling: Imagery, Shapes and Techniques*. Pittsburgh: ETC Press.

Godemann, J., & Michelsen, G. (eds.) (2011). *Sustainability Communication: Interdisciplinary Perspectives and Theoretical Foundations*. Dordrecht: Springer.

Haan, G. de (2002). Die Kernthemen der Bildung für eine nachhaltige Entwicklung [The Core Themes of Education for Sustainable Development]. *ZEP – Zeitschrift für internationale Bildungsforschung und Entwicklungspädagogik*, (1), pp. 13–20.

Herrera, F., Bailenson, J., Weisz, E., Ogle, E., & Zaki, J. (2018). Building long-term empathy: A large-scale comparison of traditional and virtual reality perspective-taking. *PloS One*, *13*(10), e0204494.

Hinyard, L. J., & Kreuter, M. W. (2007). Using narrative communication as a tool for health behavior change: A conceptual, theoretical, and empirical overview. *Health Education & Behavior*, *34*(5), pp. 777–792.

Jenkins, H. (2007). *The Wow Climax: Tracing the Emotional Impact of Popular Culture*. New York: University Press.

Kates, R. W. (2011). What kind of a science is sustainability science? *Proceedings of the National Academy of Sciences of the United States of America*, *108*(49), pp. 19449–19450.

Larkey, L. K., & Gonzalez, J. (2007). Storytelling for promoting colorectal cancer prevention and early detection among Latinos. *Patient Education and Counseling*, *67*(3 SPEC. ISS), pp. 272–278.

Lee, E., & Leets, L. (2002). Persuasive storytelling by hate groups online. *American Behavioral Scientist*, *45*(6), pp. 927–957.

Lengauer, G., Esser, F., & Berganza, R. (2012). Negativity in political news: A review of concepts, operationalizations and key findings. *Journalism: Theory, Practice & Criticism*, *13*(2), pp. 179–202.

Masschelein, J. (2004). How to conceive of critical educational theory today? *Journal of Philosophy of Education*, *38*(3), pp. 351–367.

McGreavy, B., & Lindenfeld, L. (2014). Entertaining our way to engagement? Climate change films and sustainable development values. *International Journal of Sustainable Development, 17*(2), pp. 123–136.
Micheletti, M., & Stolle, D. (2012). Sustainable citizenship and the new politics of consumption. *The Annals of the American Academy of Political and Social Science, 644*(1), pp. 88–120.
Nwoga, I. A. (2000). African American mothers use stories for family sexuality education. MCN. *The American Journal of Maternal Child Nursing, 25*(1), pp. 31–36.
Rogers, E. (2003). *Diffusion of Innovations* (5th ed.). New York: Free Press.
Rosin, M., Wong, J., O'Connell, K., Storksdieck, M., & Keys, B. (2019). Guerilla science: Mixing science with art, music and play in unusual settings. *Leonardo*, pp. 1–11.
Singhal, A. (2013). Introduction: Fairy tales to digital games: The rising tide of entertainment education. *Critical Arts, 27*(1), pp. 1–8.
Singhal, A., Wang, H., & Rogers, E. (2013). The entertainment-education communication strategy in communication campaigns. In R. E. Rice & C. K. Atkin (eds.), *Public Communication Campaigns* (4th ed., pp. 321–334). London: Sage.
Smith, J., Butler, R., Day, R. J., Goodbody, A. H., Llewellyn, D. H., Rohse, M., Smith, B. T., Tyszczuk, R. A., Udall, J., & Whyte, N. M. (2017). Gathering around stories: Interdisciplinary experiments in support of energy system transitions. *Energy Research & Social Science, 31*, pp. 284–294.
Sugiyama, M. S. (1996). On the origins of narrative: Storyteller bias as a fitness-enhancing strategy. *Human nature*, 7(4), pp. 403–425.
Tobias, R. B. (2011). *20 Master Plots: And How to Build Them* (3rd ed.). Ohio: F+W Media.
Van Hulst, M., & Ybema, S. (2020). From what to where: A setting-sensitive approach to organizational storytelling. *Organization Studies, 41*(3), pp. 365–391.
Vervoort, J. M., Keuskamp, D. H., Kok, K., van Lammeren, R., Stolk, T., Veldkamp, T., Rekveld, J., Schelfhout, R., Teklenburg, B., Cavalheiro Borges, A., Jánošková, S., Wits, W., Assmann, N., Abdi Dezfouli, E., Cunningham, K., Nordeman, B., & Rowlands, H. (2014). A sense of change: Media designers and artists communicating about complexity in social-ecological systems. *Ecology and Society, 19*(3), 10.
Wales, K. (2011). *A Dictionary of Stylistics*. London: Routledge.
Weischer, A. E., Weibler, J., & Petersen, M. (2013). "To thine own self be true": The effects of enactment and life storytelling on perceived leader authenticity. *The Leadership Quarterly, 24*(4), pp. 477–495.
Woodside, A. G. (2010). Brand-consumer storytelling theory and research: Introduction to a Psychology & Marketing special issue. *Psychology & Marketing, 27*(6), pp. 531–540.

2

DOES STORYTELLING FOR SUSTAINABILITY WORK? AN EXPERIMENT

Anna Sundermann, Daniel Fischer, Sonja Fücker, Daniel Hanss, and Hanna Selm

This chapter describes a study that explores two questions. First, what is the effect of text-based Storytelling for Sustainability (SusTelling) among young adults aged 18–24? And, second, do SusTelling effects differ as a function of young adults' level of interest in sustainability topics and educational background (i.e., higher education vs. vocational training)?

We begin by outlining the background of our study. In the next section, we describe the methods and results of a field experiment, together with the results of subsequent group discussions. Finally, we discuss the results of these two research components jointly and derive recommendations for the use of SusTelling in practice.

2.1 Great Expectations – And Nothing More?

Recently, there has been an increasing interest in promoting narrative communication and storytelling in fields such as science communication (Lidskog et al. 2020; Taddicken & Reif 2020; Kaul et al. 2020), corporate social responsibility reporting (Coombs 2019; Boukes & LaMarre 2021), and (environmental) journalism (Lück et al. 2018; Smith 2005). Behind these research interests are public expectations that storytelling will have persuasive effects on audiences.[1]

Practitioners in the respective fields often assume that storytelling is intuitive and easy to understand, arouses the interest of an audience, and is perceived as entertaining (Dahlstrom 2014). It is also expected that easily recognizable characteristics in storytelling (see Chapter 1) make this style of communication especially suitable for presenting complex issues to diverse audiences (Machill et al. 2006). However, these expectations differ depending on the specific field of practice and research. For instance, studies in the field of corporate communications investigate the effects of storytelling in advertising and communication strategy or employee

DOI: 10.4324/9781003326144-5

management (e.g., Delgado-Ballester & Fernández-Sabiote 2016; Sanders & van Krieken 2018; Woodside et al. 2008). In the fields of health, risk, and environmental communication, the focus is on the knowledge, beliefs, attitudes, and behavior of individuals (e.g., Dunlop et al. 2008; Kim et al. 2012; Lee et al. 2016). And in educational sciences, and especially in didactics, storytelling is expected to support the teaching of subject matter, making it more understandable and accessible to learners (e.g., Adams et al. 2012; Cunningham & Gall 1990; Wannagat et al. 2021).

Considering the many diverse effects, storytelling is assumed to have, it is not surprising that it is increasingly being applied in an attempt to promote sustainability as a social vision (see Chapter 1).

However, research findings on the impact of storytelling, in general, are ambiguous. An initial descriptive review of 70 studies on the effects of narrative communication[2] shows that there is only sufficient evidence for positive effects on memory, attitudes, and behavioral intentions (Frey 2014). The review furthermore presents evidence to support the following effects: a more vivid and realistic assessment of contextual details, increased expectations of self-efficacy, and the experience of specific emotions during reception. It concludes that narrative communication has *only short-term effects* on message-compliant attitudes, behavioral intentions and behaviors, and an increased experience of transportation (i.e., being drawn into the described reality) after reception. However, the authors also note that narrative communication may be inferior to nonnarrative communication regarding perceived conveyed information and short-term knowledge acquisition. Additional uncertainties regarding the effects of storytelling arise from the diversity of conceptualizations and definitions of storytelling and narrative communication (see Chapter 1). In contrast, a meta-analysis (n =14) of the persuasive effects of narrative communication presented three findings: First, short-term persuasive effects of narrative messages were found for attitudes and behavioral intentions; second, long-term effects were found for not only attitudes and intentions but also behaviors. However, the narrative and nonnarrative messages did not affect participants' beliefs differently (Oschatz & Marker 2020). Another recent meta-analysis confirmed the superiority of narrative texts compared to expository texts for content comprehension and recall (Mar et al. 2021).

We see that the current state of research leaves many questions unanswered. Is storytelling in the context of sustainability communication superior to more traditional forms of communication? And what role do sociodemographic characteristics of the recipients (such as their educational level) play here?

We conclude that there is a lack of knowledge as to the extent narrative forms of communication, such as storytelling, can activate participation in social transformation processes. Evidence-based findings could make substantial contributions to strengthening the effectiveness of sustainability communication. Against this backdrop, our field experiment addresses the questions we have just described. We supplemented the experiment with two focus group interviews among participants of the experiment, to be able to also interpret the results from their perspectives.

In the following section, we describe the methods used in the experiment.

2.2 A Field Experiment on the Effects of SusTelling

In media research, experiments are frequently used to test the effects of media messages. To test hypotheses about relationships between characteristics of messages and recipients in a controlled manner, participants are given at least two different versions of a text. The experimental design must be precisely specified and narrowly defined to investigate the effects of these texts on psychological processes or mental states of the recipients (e.g., emotional states or attitudes). The careful formulation of hypotheses is crucial to the experimental method.

Our general hypothesis in this experiment is that the SusTelling style of presentation is superior to that of the classic reporting style. There is a need for additional specifications regarding the general hypothesis. Therefore, we describe three subsequent hypotheses: first, SusTelling increases situational interest and behavioral intentions.

> Hypothesis 1: Narrative-style journalistic texts compared to reporting-style journalistic texts on sustainable consumption increase situational interest, personal consumption-related, and sustainability-related engagement intentions of young adults (18–22 years).

Regarding the specific topic and context of our experiment, Hypothesis 1 predicts: After reading a text in SusTelling style concerning packaging-free shopping, students will show greater interest in the issue, have stronger intentions to engage in sustainable consumption (such as packaging-free shopping), and be more likely to (further) engage in sustainability-related voluntary work. The next two hypotheses address the relationships between recipient characteristics and design characteristics of a journalistic text (Früh & Frey 2014; Emde et al. 2016). The second hypothesis states that a journalistic text in storytelling style on a sustainability topic has a different impact on recipients with a strong interest in sustainability issues than on people with a low interest in sustainability issues.

> Hypothesis 2: Narrative-style journalistic texts have different effects on young adults with a low compared to those with a high affinity for sustainability.

The third hypothesis states that the effectiveness of SusTelling is related to the recipient's educational background.

> Hypothesis 3: Narrative-style journalistic texts have different effects on young adults in vocational training compared to those studying at a university.

Testing these hypotheses should provide more nuanced insights into the suitability of SusTelling for hard-to-reach target groups. To prevent publication bias for positive findings, we have precisely formulated and published our hypotheses prior to data collection (https://osf.io/wa75e).

There is a need for additional considerations beyond those described in this chapter (see Sundermann et al. 2019). We briefly discuss the three most important ones below.

First, young people between the ages of 15 and 24 comprise about 16% of the world's population (United Nations 2018). Thus, young adults are moving into the focus of sustainability communication. They will have to cope with the consequences of further unsustainable development and will eventually become the leaders at the head of sustainable development initiatives. The active engagement of young adults in decision-making for sustainability is therefore a central goal of the Agenda 2030. Youth surveys indicate that there is already a high level of awareness of sustainability issues such as climate change that for many youths and young adults is associated with feelings of worry or even anxiety (Hickman et al. 2021). However, surveys also show that strong active engagement toward sustainability such as changing consumption habits is related to certain personal characteristics such as higher educational level, gender, or level of interest (Dunne & Bijwaard 2021). One of the reasons for this deep gap between awareness of the problem and actual engagement may be a lack of target group segmentation, which means that sustainability communication is insufficiently geared to the needs of certain segments of the young target group (Klöckner 2015). Our experiment, therefore, focuses on German young people aged 18 to 24 years with varying levels of interest in sustainability and different educational backgrounds.[3]

Second, young people's interest in daily newspapers has been decreasing around the globe (ASDA'A BCW 2021; Flash Eurobarometer 2021; Ichikowitz Family Foundation 2020; Working Group on Youth of the Regional Collaborative Platform for Latin America and the Caribbean 2021). At the same time, social media consumption has been increasing in this demographic. If daily newspapers want to increase their appeal to young adults, they will probably need to adjust their reporting style. Against this background, we decided to study the effects of an article that had appeared in a local daily newspaper with an adapted text version in the SusTelling style (see Sundermann et al. 2019). Even though the text style, and not the content, is the main focus of the field experiment, it is nevertheless important that the message (in reporting style or SusTelling style) is relevant to the goals of sustainable development. The original newspaper article described the successful opening of a packaging-free section in a local supermarket as the outcome of a young people (with the target group's age demographics) sustainability initiative. The reason for choosing the issue of packaging-free food products (which are also organic, regional, and fair-trade) was its comparatively high influence on the individual ecological footprint (Geiger et al. 2018). Moreover, the characters in this journalistic text could serve as blueprints for the target group in demonstrating how young people can successfully influence the sustainable development of the local food supply.

Third, in this study, we hypothesize that the presentation style (reporting vs. SusTelling) will have different effects on an individual's situational interest and (consumption-related) behavioral intentions. In the influential elaboration-likelihood theory, high situational interest is characteristic of deep and intensive cognitive processing of journalistic texts, in contrast to superficial processing, for example, in the case of disinterest (Petty & Cacioppo 1986). The processes of understanding and experiencing triggered by situational interest thus form the starting point for explaining the effects of media messages on the recipients' knowledge, attitudes, and behavioral intentions. Numerous studies show that a high level of situational interest is positively related to the acquisition of knowledge and a good recall of information (Renninger & Hidi 2011). Additionally, studies from the field of political communication indicate that situational interest makes attitude change more likely to occur (Aarts & Semetko 2003; Guo et al. 2019; De Vreese & Boomgaarden 2006). In this study, we hypothesize that a SusTelling-style text will elicit higher situational interest than a reporting-style text because its design features increase suspense and thus hold the recipient's attention more strongly (Perse 1990). Increased situational interest is thus seen as a mechanism by which SusTelling can affect the processing of a story's message.

The goal of improving sustainability communication in the field of sustainable consumption restricts which behavioral intentions can be investigated (effect variables) in the field experiment. As discussed in Chapter 1, an important goal of sustainability communication is activating social participation by young adults. The texts used in this study contain the same message, but they use different styles of presentation (reporting vs. SusTelling). The explicit message of both text versions is that packaging-free shopping is environmentally and socially beneficial, suggesting that packaging-free shopping is a way to contribute to sustainable development. The description in these texts of how young adults contribute to local sustainable development sends an implicit message that the individual engagement of young adults in protecting the environment can be successful.

It is thus reasonable to hypothesize that media messages can have an effect on both types of intentions, which is why we decided to investigate the effect of texts on the sustainable consumption intentions and the environmental engagement intentions of young adults. Sustainable consumption intentions are understood as an individual's intention to perform all acts of need satisfaction in food consumption – the acquisition, use, and disposal of food – so that the bases of ecological and social life for all current and future generations are not endangered (adapted from Geiger et al. 2018; Fischer et al. 2017). Behavioral intentions directed toward environmental protection, according to Alisat and Riemer (2015), are intentions to engage in civil actions to protect the environment. These intentions are aimed at addressing the systemic causes of environmental problems, as well as promoting environmental activism.

The field experiment was conducted in September 2019 at a university and from October to December 2019 at two vocational schools in a medium-sized city in Northern Germany. Each participant received only one of the two text versions (reporting vs. SusTelling style). The individual students in each of the 32 participating vocational school classes and the individual students in six classes at their university were randomly given one of the two text versions. Data collection was conducted using paper-based survey questionnaires in the classroom to ensure a high response rate and good data quality.

Two group discussions were held after the completion of the survey questionnaires. The aim of collecting these data was to obtain further evidence of possible effects of the SusTelling-style text, also considering the perspectives of the participants. These qualitative data supplement the quantitative data gathered in the field experiment. The vocational school and university participants in the group discussions were selected among taking part in the field experiment. A detailed description of data collection and analysis is currently in preparation. The next section describes in more detail the composition of the participant sample and the results of the experiment.

2.3 Sample and Results of the Experimental Study

2.3.1 Sample

A total of 840 students participated in the experiment; they were between 18 and 24 years old (M = 20.10, SD = 1.49). Well over half of the participants (68%) were female. Half were university students, and the other half were students in vocational training. With few exceptions, participants were native speakers of the German language. The majority possessed an upper secondary school certificate, followed by those with an intermediate secondary certificate. Only 27 participants had a lower secondary school certificate. Some participants had already earned a university degree or completed their vocational training. Only very few did not state their most recent educational qualification. The university students were, with few exceptions, students of sustainability science or business administration. The vocational students were being trained in healthcare (e.g., as dental assistants, physician assistants), sales (e.g., as real estate agents, retail salespersons), management (e.g., as insurance and finance assistants, office clerks), and agriculture and animal husbandry (e.g., as veterinary assistants, farmers).

2.3.2 Results

The presentation of the results focuses on the three hypotheses of the experimental study. First, SusTelling-style texts increase situational interest and strengthen behavioral intentions. Second, there is an interaction between participants' interest in sustainability issues and the presentation style of the texts (SusTelling vs. reporting style). And, third, there is an interaction between participants' education level and the presentation style (SusTelling vs. reporting style). In general, if a result is considered "not significant", the available data do not provide sufficient statistical support for the tested hypothesis.

No Effect of Text Style on Situational Interest and Behavioral Intentions

The first hypothesis was that a text in SusTelling style would have a positive effect on situational interest, sustainable consumption intentions, and environmental protection-related engagement intentions. Figure 2.1 presents the respective mean values for situational interest, sustainable consumption intentions, and environmental engagement intentions. According to our hypothesis, if the presentation style had a positive influence on behavioral intentions, the SusTelling-style bars (right-hand side) should be significantly higher than the reporting-style bars of the same color (left-hand side). However, the situational interest (dark gray) of participants reading the SusTelling-style text does not differ significantly from that of participants reading the text in reporting style. Contrary to the hypothesis, there appears to be no effect of the SusTelling-style text on the situational interest of the participants.

It is also evident from comparisons between the narrative and nonnarrative conditions that there are no differences in the effects on either of the behavioral intentions. This means that participants who read the text in the SusTelling style have, contrary to the hypothesis, neither have higher intentions to buy sustainably (black bars) nor higher intentions to engage in environmental protection-related activities (light gray bars).

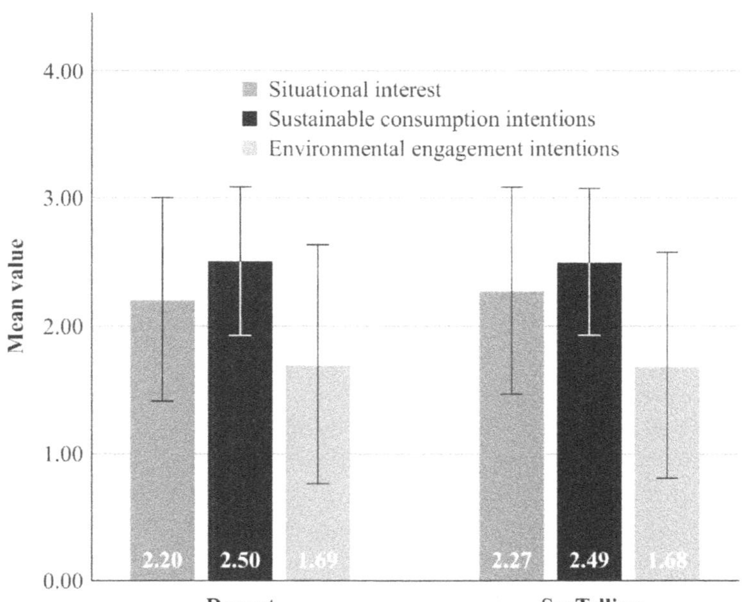

FIGURE 2.1 No statistically significant effects of the SusTelling-style text on situational interest and behavioral intentions (95% confidence interval)

No Interaction Between Interest in Sustainability and Text Style

The second hypothesis is that the SusTelling-style text would have a different effect on participants with a low interest in sustainability than on participants with high interest; that is, there is a reciprocal relationship between a participant's level of interest in sustainability and the presentation style of the text. This hypothesis was also not supported by the results. Figure 2.2 shows the respective mean values for the three dependent variables for the two versions of the text in a comparison between those with low versus high interest in sustainability. For each of the three variables, there are differences within students with low interest in sustainability (bars on the left) and within students with high interest in sustainability (bars on the right). We assumed that the differences in the dependent variables among students with low interest in sustainability between the reporting and SusTelling conditions would differ from those differences among students with high interest. However, the differences between these differences are not statistically significant. Participants interested in sustainability did show, regardless of the text version, a higher situational interest in the text as well as a stronger intention to consume sustainably and engage in environmental protection-related activities.

No Interaction Between Educational Orientation and Text Style

The third hypothesis was that the SusTelling-style text would have a different effect on students at vocational schools than those at the university. Again, an interaction

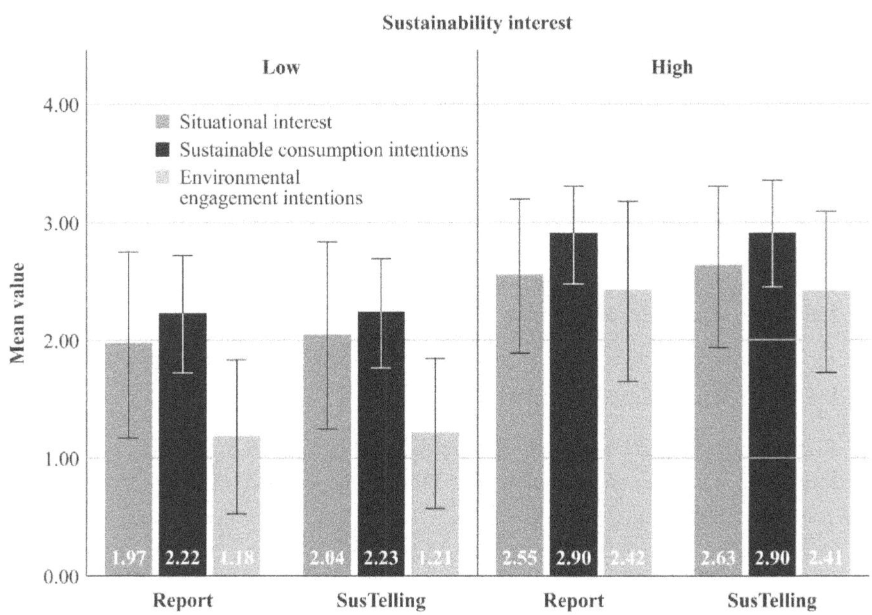

FIGURE 2.2 No statistically significant interaction between the personal characteristic of interest in sustainability issues and text style (reporting vs. SusTelling style) (95% confidence interval)

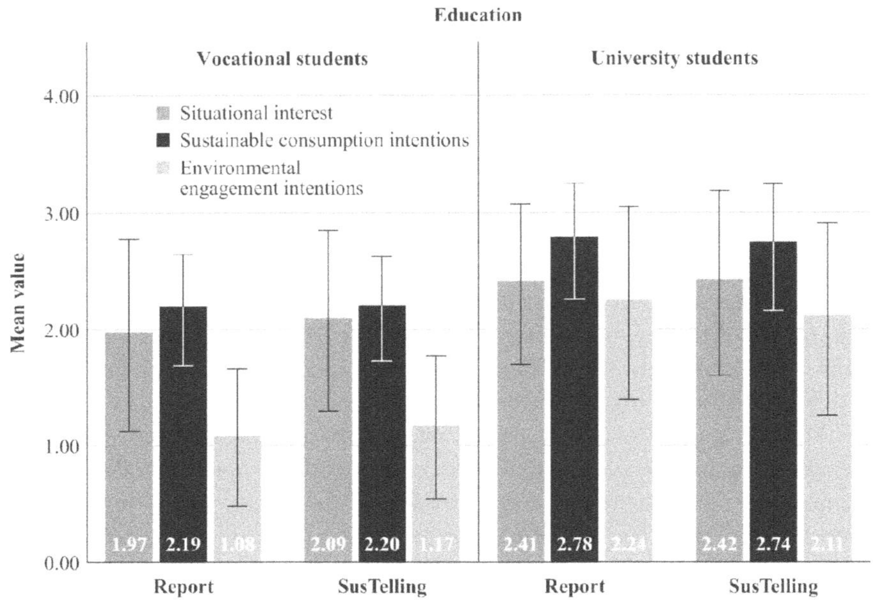

FIGURE 2.3 No statistically significant interaction between educational orientation (personal characteristic) and text style (reporting vs. SusTelling style) (95% confidence interval)

between the personal characteristic of educational orientation and the presentation style of the texts was expected, and this hypothesis could not be corroborated. Figure 2.3 shows the respective mean values for the three dependent variables separately for the two text versions in a comparison between students at the vocational school versus those at the university. For each of the three dependent variables, there should be a difference (vocational training vs. university students) in the differences (reporting vs. SusTelling style). However, the differences between the differences are not statistically significant. University students did show, regardless of the text version, a higher situational interest in the text, and a stronger intention to consume sustainably and engage in conservation-related activities.

2.4 Preliminary Conclusions

This study investigated whether a journalistic text in the SusTelling style has a greater effect on situational interest and sustainability-related behavioral intentions than a text in reporting style.

None of the hypotheses regarding the superiority of journalistic texts in the SusTelling style compared to the classic reporting style could be confirmed. Considering the high expectations in the practice of storytelling in the field of sustainability communication, these results may be surprising. However, this result is consistent with the findings of a meta-analysis by Winterbottom et al. (2008), who compared 17 studies on the impact of persuasive text messages in the field of health communication. Only one-third of the studies were the narrative presentation of texts more effective than the nonnarrative one. The results of Früh and Frey (2014) point in a similar direction. In an experimental study, the authors examined the reception effects of journalistic texts on 240 people aged 18–66 years to determine whether narrative forms of presentation are superior to nonnarrative forms in terms of comprehensibility, attractiveness, and entertainment value. The authors conclude that a hypothesis of the general superiority of narrative presentations cannot be held. They hypothesize that the effects are strongly influenced by the characteristics of the participants, the medium, and the reception situation. Overall, our results confirm Früh and Frey's (2014) conclusion that narrative formats are not fundamentally superior to reporting-style formats. On the basis of these findings and the specifics of the experimental design, we now interpret the results of the field experiment based on the characteristics of text style, recipients, and situation.

In a first interpretive step, we examined the possibility that carry-over effects due to the presentation and content of the texts influence the impact of SusTelling. The essence of the journalistic text was the opening of a packaging-free section at a local supermarket. The news item was thus about an everyday occurrence in a local context. However, Emde et al. (2016) advance the assumption that the advantages of narrative communication are particularly evident when the content has more complex interrelationships and initially includes few characters with identification potential for the target audience. It could be that SusTelling is better used for communicating complex content as it ties up less cognitive capacity, which can then be used to understand more complex subject matters. This hypothesis is further supported by a literature review on the effect of mass communication in the field of environmental guidelines (De Vries 2020). De Vries shows that understanding what is being communicated depends on both the cognitive capacity of the recipient and the complexity of the message. It would be worthwhile for follow-up studies to examine the relationship between the complexity of content and narrative presentation in the field of sustainability communication.

As already mentioned, Emde et al. (2016) also see the potential for readers to identify with characters in the text as an important factor in the effect of narrative texts. In an attempt to create identification potential for all participants in the experiment and to make the characters as neutral as possible, the three protagonists in the texts were two female and one male young adult of the target group's age. Neither of the text versions contained information about the educational background of the characters behind the initiative. However, it is relatively obvious that they were interested in sustainability, which could have appealed to those participants who were already more interested in the topic. In addition, to achieve the highest possible perceived authenticity of the texts for the participants, the message was about a local example of sustainability. However, it cannot be ruled out that the participants may have already been familiar with the initiative and the protagonists in the story, and that this influenced the reception effect. In future research, the characteristics of different female and male protagonists could be systematically varied to determine the effect they have on the identification potential for different target groups.

The second interpretive step considered situational factors that might have influenced the results. We designed the experiment explicitly to eliminate almost all situational–contextual influences. For example, the same test leaders were used and the instructions were standardized before and after the start of the survey. However, as in any field experiment, the situation in the classrooms was difficult to control. Some participants could have been distracted while reading the text and therefore engaged less intensively with its content. In future research, it would be interesting to investigate to what extent the effects of storytelling in a laboratory context differ from those in a field experiment.

Furthermore, it cannot be ruled out that the topics of plastic waste and unnecessary packaging had already been discussed in class so that a kind of saturation or fatigue effect occurred among at least some participants. This topic had been much discussed in the media so that it could have an additional negative effect on the news value of the text. It would also be interesting to consider other sustainability topics that might reflect the sociocultural dimension of sustainability (e.g., fair wages, human rights, quality of life) more strongly. It is assumed that these issues are less likely associated with sustainability among young adults (Sundermann & Fischer 2019).

Third, influences that may have been exerted by the sample and its characteristics must also be considered when interpreting the results. The study identified personal characteristics (such as educational qualification or interest in a particular topic) that have already been shown to be influential for the study of sustainability communication in other contexts. We expect that well-told stories with a schematically familiar structure require less cognitive capacity on the part of recipients, making it easier to process more complex content. We, therefore, hypothesized that the text in the storytelling style would have a different effect on the vocational students. However, it was found that a very large majority of participants reported their last educational qualification was a high school certificate. And it thus seems plausible that there is probably little difference between the two groups in terms of cognitive capacity or in familiarity with reading texts. The low variability in this personal characteristic means that there was no measurable difference in the effect of SusTelling-style text on situational interest and behavioral intentions, and that no relationship was detectable between this personal characteristic and the style of presentation.

Another effect that might have influenced the generally high scores in situational interest among interested students is known as the *echo chamber effect* (Levy & Razin 2019). This effect refers to an individual showing a higher level of agreement with content that corresponds to their views, values, and norms. Media reports that are contrary to one's worldview or interests, on the other hand, are more likely to be rejected. Since in this sample the interest in sustainability is related to the university students' studies, it could be that the more interested students had a high situational interest in the content of the text, regardless of how it was presented. The generally higher engagement intentions of university students are supported by the assumption that they are in a setting in which there is a strong social norm for environmental engagement. In addition, the characteristics of a message have an influence on perceived trust and behavioral intentions related to the message when the message design indicates that it is intended to inform rather than persuade (Rabinovich et al. 2012). The authors show that trust in information and the intention to act in a more environmentally friendly way are greater when recipients are convinced that the researchers wanted to provide information about climate change.

In light of these interpretations, it seems advisable not to assume that young adults are only enthusiastic about and pay attention to the news because it is told as a story. Based on our research findings, a closer look at the characteristics of the target group could lead to more effective sustainability communication. It also seems worthwhile to pay more attention to the selection and characteristics of the protagonists in the story and to emphasize the informative character of the message. To investigate more into these theoretical considerations from the participants' perspectives, two focus group discussions were conducted in addition to the experiment. The methods and results are presented in the following section.

2.5 In-depth Exploratory Focus Group Discussions

The goal of the two group discussions was to gain an in-depth insight into the participants' direct perceptions of the SusTelling text presented in the field experiment. To ensure comparability between the interviews, a guideline was developed that contained the same questions for all participants. At the same time, interviewers were allowed to ask follow-up questions to remain flexible in conducting the interview. In the first part of the interview, repeating and paraphrasing questions were used to evoke participants' memories of the data collection situation and their initial perceptions of the SusTelling text. This procedure was intended to provide information on the extent to which the participants could recall the text and the message. In the second part of the interview, the text in SusTelling style was handed out to the participants. By asking specific questions about their perception of its narrative form and content, more in-depth insights were gathered into the participants' impressions of the text.

To ensure a high degree of diversity among the interviewees, participants were selected on the basis of demographic data collected in the field experiment. All interview participants gave their active consent in accordance with data protection regulations (see Sundermann et al. 2019). Participants were recruited for the group discussions by directly approaching teachers and students taking part in the experimental study. One group discussion was conducted with four university students studying environmental sciences and the second with seven vocational students from the retail sector. The interviews were transcribed in their entirety and then coded for data analysis and summary using a two-part coding scheme. The first part of the coding scheme was derived from the definition of SusTelling (see Chapter 1) to identify textual features that the participants had mentioned earlier in the interview. The second part of the coding scheme was based on a literature review of the specific effects of storytelling to identify indications of potential effects and perceptions of the SusTelling-style text in the participants' comments.

2.5.1 Different Emphases in Perceptions of SusTelling

In the interviews, both groups frequently focused on the topic of sustainable consumption and frequently mentioned the option of buying package-free goods. The vocational students generally expressed a positive attitude toward the topic and also showed they were aware of the problem. However, it is apparent from their answers that the text was not newsworthy, "because there was no new information in it for me" (Gri12OI). Other vocational students expressed annoyance because "either you separate trash or you don't separate it. . . . I actually think it's important, but to get involved myself. No, thanks. I'd rather read a book" (Pac29Yes) or "Why talk about it, just do it" (Ham20Ar). In addition, the vocational students indicated in the interview that they were no longer interested in the topic of plastic waste, as they had already discussed its ecological consequences in class. One student first mentioned how his class had already dealt with the topic of plastic waste and then concluded with the words "We just somehow lost interest" (Gri12Ol). This perceived saturation with the topic of plastic waste may have led, in the experiment, to a low situational interest in the text, regardless of the text version.

The university students also described the topic as a "gigantic topic, which is, after all, somehow an issue at every level of our daily lives" (Ste30KI). However, their statements reveal a much greater enthusiasm for the topic. For example, one student said: "I think what I also really like about it is that, simply said, it's a topic I like" (Sch07Me). And another participant added: "So for me right now such a big topic is, generally speaking, regenerative agriculture" (Lie10Jo). In their statements about the topic, the students also showed a broad range of background knowledge, for example, that even with food from Europe, the working conditions are probably "not that great now either" (Lie10Jo). The students were thus not only enthusiastic about the topic but also added their interests and new facts to the text during the discussion. Overall, the biggest difference in the responses of the two groups was in their enthusiasm and interest.

The storytelling style of the news article was perceived by the vocational students as "like a novel" (Gri12OI), "not really factual" (Gri12OI), and "easy to understand" (Pac29Yes). One person felt offended by the text, "They think we're stupid!" (Pac29Ja). They also criticized that there was "far too much beating around the bush" (Pod22Li) and that it was "unnecessarily detailed" (Gril2OI). At the same time, one person stated about personalization: "I'm supposed to sympathize with the protagonists, but I just find them simply uninteresting" (Pac29Yes). It was also pointed out that it was not possible to get caught up in the story because a "climax" (Kac25Ma) was missing. However, the vocational school students were not unanimous in their evaluation of the presentation style. Some also stated that they also found its "novel-like" and "simple" qualities to be "pleasant" (Ham20AR) and could imagine the text being "something for younger people" (Gri12OI). Two people suggested that a "mixture" (Pod22Li), possibly a "newspaper article" (Gri12OI) between "novel-like" and "factual", might be the solution.

The university students found many positive terms for the stylistics, such as "descriptive" (Teu06Ma), "graphic" (Ste30Kl), "vivid" (Lie10Jo), or that a "clear picture was painted" (Ste30Kl). They described the structure of the story with a beginning and an end as well as the main characters. Several times they also positively emphasized that it was a "specific, regional example" (Ste30KI). One student emphasized the choice of words like "fresh, light green, big blue eyes" (Teu06Ma) created a positive mood. Regarding the characters, the students noticed that they were chosen "across generations" (Sch07Me) and that it was "very personally written" (Ste30Kl). At the same time, one student was struck by the fact that it was about "people of our age, and they had an idea, and they presented their idea to the right people at the right time" (Sch07Me). Unlike the vocational students, the university students did not comment on the dramaturgy.

Both groups frequently expressed their perception that the text contained a normative message of an intention to change. However, not all vocational students agreed on whether the text contained a message. For example, one criticized "that there aren't any real messages" (Pod22li), while another countered that "Shop sustainably because that's cool" was the message (Pac29Ja). Yet another vocational student added that the "why", a justification of the call for plastic-free shopping, was missing (Gri12OI). The university students took the message more uncritically as a "solution approach" (Lie10Jo, Ste30KI), which was not "overly pushy" (Teu06Ma). They criticized the text on two points: there was no discussion of the high price of the package-free products, and they perceived there was misinformation about the assortment. These comments show that the participants already had a high level of prior knowledge about the story. We can conclude that both the vocational students and the university students recognized the simple, explicit message to "shop sustainably". Beyond this simple message, the university students also seemed to perceive a second, more implicit message. They identified the message of the text as an "incentive for their projects" which means to encourage and "motivate people" in their initiatives toward sustainability (Ste30KI).

2.5.2 Different Perceptions = Different Effects?

The vocational students expressed themselves fairly negatively about the way the SusTelling-style text was presented. An informational or factual text that "simply states the important aspects" would have been much more pleasant to read, according to one student (Gri12OI). One vocational student stated, "so in my spare time now this would have been a story I wouldn't keep reading" (Ham20AR). The text was described by another as "uninteresting" because it did not "speak to the student personally" (Pac29Yes). One vocational student said that there was nothing that "grabbed" him (Kac25Ma). The style of presentation also elicited negative attitudes from one student because he felt he was being treated as "stupid" (Pac29Yes).

Overall, these statements indicate that the style of presentation elicited fairly negative emotions and attitudes among the vocational students.

The emotional and affective effects identified in the statements of the university students differed greatly from those of the vocational students. The text had put them in "such a good mood" (Lie10Jo) or made them feel "relaxed" (Teu06Ma). One student also stated that she had "reflected on it a bit more" (Teu06Ma). One student reported, "I then kind of linked it to another thing I had already been thinking about" (Sch07Me).

At the same time, it is apparent that the identification potential was high for the university student sample. One student described,

> But there are simply three people my age, and they are involved in an initiative [in the story], I'm also involved in an initiative. And that's it, I also want to take away the fear of people who are dealing with sustainability problems that are supposedly so complex.
>
> *(Scho07Me)*

Above all, the examples and comments from the interviewees clearly show how different the perception of the SusTelling-style text was in the two groups. It is also interesting to note that the vocational students learning to work in the retail trade discussed how they would inform their customers about environmental protection-related measures in the store. They expressed discomfort with the SusTelling-style text and indicated that a text or a poster would be more appropriate to inform customers than the text example.

Another important difference between the two groups was the discussion of the emotional and motivational effects of the SusTelling-style text. The university student group made more statements suggesting a motivational effect than the group of vocational students. In addition, the text triggered contrasting attitudes in the two groups: mostly positive attitudes among the university students and more negative ones among the vocational students. Two university students described effects that indicate a strong identification with the protagonists in the text. This type of message effect was not mentioned by the vocational students, which suggests that the characteristics of the protagonists are crucial for the target audience to identify with the content.

We can conclude that the vocational students tended to have a negative attitude toward the topic and the way it was presented. The university students were largely enthusiastic about the nature of the initiative and had very positive attitudes toward the design features of the text. Contrary to initial expectations, the text appeared to offer more identification potential for the university students than for the vocational students. The text content and design thus seem ill-suited to motivate vocational students from the retail sector to participate in shaping a sustainable future, or even to show interest in the topic of package-free shopping.

Given the results of the field experiment and the group interviews, we now present recommendations for the application of SusTelling in practice.

2.6 Recommendations

Looking at the results of both the field experiment and the group discussions, it becomes apparent that future research about the effectiveness of storytelling in sustainability communication will have to be much more nuanced. At least in the field experiment, reported here, we observed that a message with a realistic theme, strong positive tonality, and protagonists committed to package-free shopping did not affect situational interest and behavioral intentions. Furthermore, we observed no interactions of the personal characteristics "interest in sustainability" or "educational background" with the style of text presentation.

The results of the discussions provide some indication of how the text material was perceived by the participants. While the university students of Environmental Science perceived the SusTelling-style text as positive and affirming, the vocational students from the retail sector were skeptical and generally negative about the text. To create a stronger appeal for this more skeptical target group, three aspects are seemingly particularly important. There needs to be, first, a balance between factual information and narrative presentation, second, protagonists with sufficient identification potential for the target group, and, third, transparency about the communication goal.

Looking at the results of both studies together, situational interest in connection with personal interest or disinterest turns out to be one of the central relationships of SusTelling. It, therefore, seems essential to consider the characteristics of the recipients for indications about how to create a SusTelling story.

Above all, however, this study shows how great the need is for further research on SusTelling. Experiments always face multiple limitations. In this study, only one text was presented in two different versions. However, given that a variety of other communication channels such as videos or podcasts are available, this is a simplified view of the possible uses of SusTelling. In any case, future research should systematically examine the gaps identified in this study to make the use of SusTelling more effective in reaching the crucial target group of young adults.

Notes

1 By "effects" we mean all actual or possible changes of any conceivable factor in individuals that are directly or indirectly related to a journalistic text and are attributable to its influence.
2 Früh and Frey (2014) use the term "narrative communication" as a distinct form of storytelling. This term encompasses all kinds of communicative formats that show a gradually differing level of narrativity or have been labeled as stories, narrations, narratives, or narrative communications.
3 Originally, the study focused on young adults aged 18–22. The target group had to be extended to 18–24 to reach a sufficient sample size.

References

Aarts, K., & Semetko, H. A. (2003). The divided electorate: Media use and political involvement. *The Journal of Politics*, *65*(3), pp. 759–784.

Adams, D. M., Mayer, R. E., MacNamara, A., Koenig, A., & Wainess, R. (2012). Narrative games for learning: Testing the discovery and narrative hypotheses. *Journal of Educational Psychology, 104*(1), pp. 235–249.

Alisat, S., & Riemer, M. (2015). The environmental action scale: Development and psychometric evaluation. *Journal of Environmental Psychology, 43*, pp. 13–23.

ASDA'A BCW (2021). *13th Annual ASDA'A BCW Arab Youth Survey*. [https://bit.ly/3t1smy2]. Accessed on May 20, 2022.

Boukes, M., & LaMarre, H. L. (2021). Narrative persuasion by corporate CSR messages: The impact of narrative richness on attitudes and behavioral intentions via character identification, transportation, and message credibility. *Public Relations Review, 47*(5), 102107.

Coombs, T. (2019). Transmedia storytelling: A potentially vital resource for CSR communication. *Corporate Communications: An International Journal, 24*(2), pp. 351–367.

Cunningham, L. J., & Gall, M. D. (1990). The effects of expository and narrative prose on student achievement and attitudes toward textbooks. *The Journal of Experimental Education, 58*(3), pp. 165–175

Dahlstrom, M. F. (2014). Using narratives and storytelling to communicate science with nonexpert audiences. *Proceedings of the National Academy of Sciences, 111*(Suppl. 4), pp. 13614–13620.

Delgado-Ballester, E., & Fernández-Sabiote, E. (2016). "Once upon a brand": Storytelling practices by Spanish brands. *Spanish Journal of Marketing-ESIC, 20*(2), pp. 115–131.

De Vreese, C. H., & Boomgaarden, H. (2006). News, political knowledge and participation: The differential effects of news media exposure on political knowledge and participation. *Acta Politica, 41*(4), pp. 317–341.

De Vries, G. (2020). Public communication as a tool to implement environmental policies. *Social Issues and Policy Review, 14*(1), pp. 244–272.

Dunlop, S., Wakefield, M., & Kashima, Y. (2008). Can you feel it? Negative emotion, risk, and narrative in health communication. *Media Psychology, 11*(1), pp. 52–75.

Dunne, A., & Bijwaard, D. (2021). *Pan-European Survey. Main multi-country report*. [https://bit.ly/3wlMVal]. Accessed on August 22, 2022.

Emde, K., Klimmt, C., & Schluetz, D. M. (2016). Does storytelling help adolescents to process the news? A comparison of narrative news and the inverted pyramid. *Journalism Studies, 17*(5), pp. 608–627.

European Parliament (2021). *Flash Eurobarometer. European Parliament Youth Survey*. [https://bit.ly/3MYZcHV]. Accessed on May 20, 2022.

Fischer, D., Böhme, T., & Geiger, S. M. (2017). Measuring young consumers' sustainable consumption behavior: Development and validation of the YCSCB scale. *Young Consumers, 18*(3), pp. 312–326.

Frey, F. (2014). Wirkungen des Narrativen. Ein systematischer Forschungsüberblick zu Effekten narrativer Kommunikationsformen [Effects of narratives. A systematic research overview of effects of narrative communication forms]. In W. Früh & F. Frey (eds.), *Narration und Storytelling. Theorie und empirische Befunde* [Narration and Storytelling: Theory and Empirical Evidence] (pp. 120–192). Cologne: Herbert von Halem Verlag.

Früh, W., & Frey, F. (2014). Positive Effekte von Narration und Storytelling: Attraktivität, Verständlichkeit und Unterhaltung. Eine empirische Untersuchung zu deren Stärke und Geltungsbereich [Positive effects of narration and storytelling: Appeal, comprehensibility, and entertainment. An empirical investigation into their strength and scope]. In W. Früh & F. Frey (eds.), *Narration und Storytelling. Theorie und empirische Befunde* [Narration and Storytelling: Theory and Empirical Evidence] (pp. 262–318). Cologne: Herbert von Halem Verlag.

Geiger, S. M., Fischer, D., & Schrader, U. (2018). Measuring what matters in sustainable consumption: An integrative framework for the selection of relevant behaviors. *Sustainable Development, 26*(1), pp. 18–33.

Guo, L., Su, C., & Lee, H. (2019). Effects of issue involvement, news attention, perceived knowledge, and perceived influence of anti-corruption news on Chinese students' political participation. *Journalism & Mass Communication Quarterly, 96*(2), pp. 452–472.

Hickman, C., Marks, E., Pihkala, P., Clayton, S., Lewandowski, R. E., Mayall, E. E., Wray, B., Mellor, C., & van Susteren, L. (2021). Climate anxiety in children and young people and their beliefs about government responses to climate change: A global survey. *The Lancet Planetary Health, 5*(12), pp. e863–e873.

Ichikowitz Family Foundation (2020). *African Youth Survey*. [https://bit.ly/3ah1mnO]. Accessed on May 20, 2022.

Kaul, L., Schrögel, P., & Humm, C. (2020). Environmental science communication for a young audience: A case study on the #EarthOvershootDay campaign on YouTube. *Frontiers in Communication, 5*, Article 601177.

Kim, H. S., Bigman, C. A., Leader, A. E., Lerman, C., & Cappella, J. N. (2012). Narrative health communication and behavior change: The influence of exemplars in the news on intention to quit smoking. *Journal of Communication, 62*(3), pp. 473–492.

Klöckner, C. A. (2015). Target group segmentation – Why knowing your audience is important. In *The Psychology of Pro-Environmental Communication* (pp. 146–160). London: Palgrave Macmillan.

Lee, H., Fawcett, J., & DeMarco, R. (2016). Storytelling/narrative theory to address health communication with minority populations. *Applied Nursing Research, 30*, pp. 58–60.

Levy, G., & Razin, R. (2019). Echo chambers and their effects on economic and political outcomes. *Annual Review of Economics, 11*, pp. 303–328.

Lidskog, R., Berg, M., Gustafsson, K. M., & Löfmarck, E. (2020). Cold science meets hot weather: Environmental threats, emotional messages and scientific storytelling. *Media and Communication, 8*(1), pp. 118–128.

Lück, J., Wessler, H., Wozniak, A., & Lycarião, D. (2018). Counterbalancing global media frames with nationally colored narratives: A comparative study of news narratives and news framing in the climate change coverage of five countries. *Journalism, 19*(12), pp. 1635–1656.

Machill, M. Köhler, S., & Waldhauser, M. (2006). Narrative Fernsehnachrichten. Ein Experiment zur Innovation journalistischer Darstellungsformen. [Narrative television news. An experiment on the innovation of journalistic forms of presentation.] *Publizistik, 51*(4), pp. 479–497.

Mar, R. A., Li, J., Nguyen, A. T. P., & Ta, C. P. (2021). Memory and comprehension of narrative versus expository texts: A meta-analysis. *Psychonomic Bulletin & Review, 28*(3), pp. 732–749.

Oschatz, C., & Marker, C. (2020). Long-term persuasive effects in narrative communication research: A meta-analysis. *Journal of Communication, 70*(4), pp. 473–496.

Perse, E. M. (1990). Involvement with local television news cognitive and emotional dimensions. *Human Communication Research, 16*(4), pp. 556–581.

Petty, R. E., & Cacioppo, J. T. (1986). The elaboration likelihood model of persuasion. In R. E. Petty & J. T. Cacioppo (eds.), *Communication and Persuasion: Central and Peripheral Routes to Attitude Change* (pp. 1–24). New York: Springer.

Rabinovich, A., Morton, T. A., Postmes, T., & Verplanken, B. (2012). Collective self and individual choice: The effects of inter-group comparative context on environmental values and behaviour. *British Journal of Social Psychology, 51*(4), pp. 551–569.

Renninger, K. A., & Hidi, S. (2011). Revisiting the conceptualization, measurement and generation of interest. *Educational Psychologist*, *46*(3), pp. 168–184.

Sanders, J., & van Krieken, K. (2018). Exploring narrative structure and hero enactment in brand stories. *Frontiers in Psychology*, *9*, 1645.

Smith, J. (2005). Dangerous news: Media decision making about climate change risk. *Risk Analysis*, *25*(6), pp. 1471–1482.

Sundermann, A., & Fischer, D. (2019). How does sustainability become professionally relevant? Exploring the role of sustainability conceptions in first year students. *Sustainability*, *11*(19), 5155.

Sundermann, A., Fischer, D., Selm, H., Silver, J., & Fischer, S. (2019). *Technical Report on the Experimental Study of the SusTelling Project*. [https://bit.ly/3QYZzE3]. Accessed on August 22, 2022.

Taddicken, M., & Reif, A. (2020). Between evidence and emotions: Emotional appeals in science communication. *Media and Communication*, *8*(1), pp. 101–106.

United Nations. (2019). *World Youth Report 2018: Youth and the 2030 Agenda for Sustainable Development*. New York: United Nations.

Wannagat, W., Steinicke, V., Tibken, C., & Nieding, G. (2021). Same topic, different genre: Elementary school children's mental representations of information embedded in narrative and expository texts. *Learning and Instruction*, 101559.

Winterbottom, A., Bekker, H. L., Conner, M., & Mooney, A. (2008). Does narrative information bias individual's decision making? A systematic review. *Social Science & Medicine*, *67*(12), pp. 2079–2088.

Woodside, A. G., Sood, S., & Miller, K. E. (2008). When consumers and brands talk: Storytelling theory and research in psychology and marketing. *Psychology & Marketing*, *25*(2), pp. 97–145.

Working Group on Youth of the Regional Collaborative Platform for Latin America and the Caribbean (2021). *United Nations Survey on Latin American and Caribbean Youth within the Context of the COVID-19 Pandemic (LC/TS.2021/68)*. Santiago: United Nations.

3

SUSTELLING IN THE RECEPTION LAB – ON HOW YOUNG ADULTS PERCEIVE STORYTELLING FOR SUSTAINABILITY

Sonja Fücker, Daniel Fischer, Hanna Selm, and Anna Sundermann

This chapter takes a look at how Storytelling for Sustainability (SusTelling) is perceived and interpreted by young adults. The empirical results from an explorative study shed light on *what* is considered important in the reception of narrative sustainability communication and *how* young adults between the ages of 18 and 24 engage with narrative content. Practitioners should gain ideas as to what building blocks are needed for successful SusTelling and how narrative content and techniques can be used to reach young adults, an important target group in the field of sustainable development. This chapter also shows the problem of studies of this kind in reaching target groups outside a particular educational background. This chapter concludes with an outlook on what should be put on the agenda in both the research and practice of SusTelling.

3.1 Narrative Sustainability Communication for Young Adults

Young adults are particularly affected by the sociopolitical discussion on sustainable development. They and future generations are the ones who will be especially affected by global changes brought about by climate change being caused today and in the past. This makes it all the more necessary to strengthen their awareness of global environmental changes and their willingness to help shape a much-needed change toward social and environmental sustainability. Practitioners from various communication fields are increasingly called upon to participate in this educational mission. They are faced with the challenge that they should convey complex interrelationships and abstract information about the extent to which the natural environment is being destroyed in a way that is appropriate for the media but at the same time encourage their audience to participate in the transformation to a sustainable society. The challenge this mission presents becomes all the clearer with

DOI: 10.4324/9781003326144-6

a look at the research literature: Interested members of the younger generation are dissatisfied with what environmental communication has to offer and want sustainability issues to be better communicated. Above all, they would like to learn more about climate protection, the energy transition, and the promotion of climate-friendly lifestyles (Fischer et al. 2017). At the same time, young adults perceive environmental education in school as insufficient in terms of quantity and lacking in offerings that are "interestingly prepared, solution-oriented and competence-strengthening" and that have a motivating effect on their sustainability actions and behavior (Gossen et al. 2016, p. 22).

Although young adults have a high affinity for social and digital media, traditional channels of environmental communication in the mass media and the Internet are only partially successful in reaching this audience (Gossen et al. 2016). The use of digital media has become everyday normality for young adults (Gossen et al. 2016, p. 18), and almost every young person between 16 and 29 uses the Internet daily (Eurostat 2020). Their search for issues concerns them, such as sustainability, the environment, and climate, via Google and via social media such as social networks or YouTube (Hodson et al. 2020). Importantly, there is a preference for image- or video-based content (BMU 2018). At this interface between the need for information, media preferences, and the demand for better and more interesting communication formats, sustainability communication will have to develop creative and diverse methods to meet the needs of young adults.

Storytelling is considered a promising tool in environmental and sustainability communication. And especially in the context of serious storytelling (Lugmayr et al. 2016), it is thought to have the potential to make socio-ecological harms tangible (see Chapter 1). Narrative formats in the form of *cross-media storytelling* (Kolodzy 2013) can be disseminated in very different channels such as text, video, image, or audio media. Digital storytelling formats such as video portals or podcasts are seen as having the potential to reach marginalized and educationally disadvantaged groups (De Jager et al. 2017). With a view to the potential of narrated sustainability communication discussed in the academic literature, this chapter aims to answer the question of how SusTelling in different media formats is experienced by young adults. For this purpose, an explorative reception study was conducted to investigate how young adults perceive and interpret SusTelling. Which content-related, stylistic, or narrative features do the target group focus on? What do they consider important or unimportant in its reception? And what do they consider ingredients for effective sustainability narratives?

To answer these questions, the study used various media formats (text, audio, video, and image) in different fields of sustainable consumption – nutrition, construction/housing, tourism, clothing, and mobility – as research material. The topic area of sustainable consumption was selected for the study, based on empirical evidence that it is of particular interest to the target group of young people (Fischer et al. 2017). A diverse selection of topics and media was used to capture the widest possible range of media usage among young adults and so gain both a comprehensive and an in-depth understanding of their reception behavior.[1] As a final step in the analysis, a reception typology was formed from the data to provide practitioners with ideas for a target group-based approach to SusTelling.

3.2 The Study – On the Reception of SusTelling

3.2.1 Methodology

The target group was young adults with different educational qualifications between the ages of 18 and 24 years. The participants were largely drawn from the vocational school students and college students participating in the experimental study (see Chapter 2). Additional participants in the study were found through network partners and project experts, the project advisory board, as well as through social media channels, and the project newsletter.

The composition of the target group provides a first insight into the data. The study involved 139 respondents (14% male, 61% female, 25% not specified) between the ages of 18 and 24. Sixty percent of the participants had a (vocational) high school diploma as their highest level of education, while 4% already had an academic degree and 11% had a lower secondary school diploma, a middle school diploma, or had completed vocational training. A quarter of respondents (25%) did not state their level of education. The homogeneous composition of the target group is also reflected in research on storytelling approaches. Participants in studies like these are predominantly young and female and come from middle-class milieus with high levels of formal education (Frey 2014, p. 141). With this sample in our study, we were unable to investigate the possible influence of different educational backgrounds on the reception behavior of the participants as we originally intended. The problem of one-sided sampling with more highly educated young adults is explained by survey research examining the relationship between environmental awareness and formal educational levels. These studies show that formal education levels determine whether and to what extent there is interest in environmental issues and sustainability, and thus who participates in studies of this kind. In a representative study of environmental awareness among 14- and 25-year-olds in Germany, 37% were actively interested in sustainability. These individuals were "sustainability-oriented" (15%) and "environmentally concerned" (22%). Those who were "environmentally passive" (37%) made up a comparable proportion of the two types, open to ecological issues and acting. Seventy percent of the young generation, based on 1,500 respondents to the Greenpeace Sustainability Barometer, are considered to subscribe to sustainability-related values, with different motivations for aligning their actions with such values (Michelsen et al. 2016; see also Kress (2021) for a summary of the more recent survey in English). Milieu-related differences have a clear impact on this distribution, as the educational background has a significant influence on interest in sustainability issues (Stevenson & Peterson 2016). While young people with an affinity for sustainability hotspot topics like climate change belong to upper-middle-class milieus, which have high social, cultural, and economic capital, the group of environmental critics or passives comes from social milieus with a low level of formal education and low social and economic status (Weckroth & Ala-Mantila 2022).

During the period from February to August 2019, participants received media messages bi-weekly in text, video, image, or audio format on selected topics related to sustainable consumption. A wide variety of 50 media messages – ranging from life in a tiny house to bicycle mobility and cheap clothing – were sent via a personalized online link and could be accessed on the online blog set up specifically for the study. The sending and publication on the online blog of the messages took place by random selection. This procedure was chosen to enable the most varied reception possible concerning subject areas and the type of medium. In the sense of a "multimodal media reception" (Bucher 2017), it was assumed that recipients subjectively decide which topic in which medium they turn to and how, thus influencing the frequency and intensity of reception. In other words, participants are not equally interested in all topics relating to sustainable consumption, and certain media are received preferentially. Such selection processes provide information on whether a particular media message arouses interest and whether its reception is considered worthwhile (added value or utility) (Krcmar & Strizhakova 2009; Levy & Windahl 1985). SusTelling characteristics were used as criteria for the selection of media messages (see Chapter 1).

The study asked how young adults engage with SusTelling-style media content, that is, how they relate to protagonists and their experiences in sustainability narratives. Research on narrative communication has shown that such confrontations occur through identification with characters and perceived similarity, that is, through subjective perceptions and interpretations of the recipients based on their social and cultural lifeworld (Moyer-Gusé 2008).

To collect the data of interest, sociodemographic data such as age, educational attainment, German language proficiency, and gender were first obtained through a questionnaire survey. This information was supplemented by inquiries about habits of using various media and about interest in or prior knowledge of sustainability issues. As a further component of data collection, a comment function set up on the online blog enabled participants to leave comments, notes, and interpretations regarding the storytelling items. Throughout the study, 36 of the 139 total participants posted 250 comments on the 50 media messages. Of the 36 commenters, 30 had a (vocational) high school diploma or a university degree as their highest educational qualification. In addition to the analysis of the comment texts, seven problem-centered interviews were conducted after the initial communication act (post-communicatively) with participants who also used the comment function. This step of data collection was aimed at gaining deeper insights into the interpretations participants formulated in the comment texts. Like the commentary texts, the interview transcripts were analyzed using qualitative content analysis (Kuckartz 2014; Schreier 2012) with the analysis software MaxQDA.

3.2.2 Empirical Results

The results show us, in a first step, which characteristics are important elements of SusTelling for the young adults in this study. The reception data collected in interviews and comments shed light on which narrative features arouse interest among young adults, how they make topics and content more comprehensible, and where the limits of storytelling lie. Our discussion of the results focuses on how young adults engage with SusTelling, that is, how the participants in the study engage with the storytelling messages and what the differences are regarding media type.

True to Life, Plausible, and Differentiated – What Makes Storytelling Successful

The results of the study show that young adults who participated in the study find those stories appealing that connect to their lifeworld, prove to be credible and factual, and give the impression that topics and problems are viewed in diverse ways from different and contrasting perspectives.

SusTelling Should Be Authentic and Experiential

Narrative content is perceived as successful by the participants when their subjective experiences can be connected with the sustainability narratives, which they can "somehow still integrate into their own life" and "use" (oso5zt4) in everyday life. Recipients find it important that a story is "well-told" (pa86o5m), "plausible, easy to imagine" (iwi6q43), contains "real-life narratives" (f0gsnd5), and is thus "related to everyday life" (mooj98o). Participants describe wanting to be offered "new interesting insights" through stories that are "relevant to their own life, practical and instructive" (pa86o5m) and thus have personal identification value. Identification processes of this kind take place using narrative techniques of personalization and are an important tool for representing information in narratives (Reisch & Bietz 2011, see also Chapter 1). With a look at the research, our findings thus provide further evidence that the potential for identifying with characters is an integral part of narrative forms of communication (Cohen 2001; Moyer-Gusé 2008) and is directly related to the stylistic device of personalization (Dahlstrom 2014). By allowing recipients to identify with characters and their experiences, they can relate cognitive and/or emotional aspects of knowledge to the narrated content and remember it better:

> That . . . if, for example, someone has talked about their life, it's immediately something that stays in your memory, because you find it so, yes, because you just find it great, so like, yes, simply exciting.
>
> *(tvjv5ev)*

One participant also describes the importance of stylistic devices of personalization in storytelling formats in contrast to more factual formats:

> So, if someone now tells me something completely new that I've never had anything to do with it, that's of course also exciting, but I think to be able to process it properly, I somehow still need something I can relate to, where I can start.
>
> *(oefozve)*

"Something relatable" is usually created by making it tangible. Storytelling, in the view of the participants, makes abstract and complex information more tangible by "giving it a voice" through characters and their worlds of experience (qb4o4b0). Stories enable recipients to "empathize" (qb4o4b0) with topics and protagonists because "you can kind of emotionally understand why they're doing what they're doing or what led them to do it" (pa86o5m). In the view of the participants in our study, the possibility of identifying with characters in narratives depends on whether one is "personally touched" and can "somehow relate to a character, so whether it's just that they have a similar background, or are the same gender" (oefozve). One interviewee describes this process as follows:

> I often only hear numbers and bad things, but I can't build a real relationship with what I hear. So, when I hear that so-and-so many people died in the accident or because of this and this catastrophe, then I can't build up a real relationship. . . . And if you hear individual stories, I think it's easier to relate to them.
>
> *(oefozve)*

Biographical stories, experiences of pioneering achievements, or creative ideas of sustainability that are communicated through characters in narratives enable recipients to slip into their roles. The experiences of characters are thus "representative of other people" (oefozve) and make information, contexts, and situations tangible. Storytelling messages that do let the recipients identify with characters regularly receive negative evaluations: "The text is good because it is written very personally, however, the author annoys me somewhat with her fear . . . which is why I find this story rather unnecessary" (fz98c93). The role of personalization as a stylistic device also points to the importance of concern or involvement in storytelling. In narrative communication, involvement offers information about the extent of a recipient's subjective motivation to engage with the content of narratives. And the connection between a message conveyed by narrative elements and its meaning for one's own life becomes clear (Levy & Windahl 1985, p. 11). Psychologically, the ability to identify with characters is influenced by a state of tension arising from a feeling of participating in their fate (Zillmann 1996). Through this kind of participation, and the experience of agency, processes are triggered in recipients that can inspire or motivate them to adopt the role and perspective of narrative characters:

> If you are presented with a person who is doing this, or who is developing this, and then you experience this whole process, I think you can clearly see, okay, then I could do this too, or I could also do something at least like this in the same way.
>
> *(oso5zt4)*

SusTelling Should Be Plausible and Realistic

For the recipients in the study, it is also important that sustainability-based narrative elements are portrayed as realistically as possible. A skillful balancing of narrative techniques and objective or factual information is essential. A harmonious composition of these two components requires "a through-line", making a narrative "coherent in itself" so that recipients do not "have the feeling that information is left out somewhere, or things are not explained enough" (qb4o4b0). According to one recipient, stories should be "told in a balanced way . . . and not exaggerated" and they should "somehow not make things heroic" (pa86o5m). Stories are perceived by him as being "clearly more exciting . . . if you somehow know, yes, this happened in the real world" (pa86o5m). Stories must meet the standard of "internal plausibility" (Esposito 2007, p. 14) so that recipients perceive narratives as likely. Sustainability narratives, from the perspective of participants, must perform a balancing act of generating interest and suspense through the use of typical stylistic devices (see Chapter 1), while providing authentic information that reveals a coherent fact-based narrative, so fulfilling the educational mission of sustainable development.

Participants in the study negatively evaluated storytelling contributions in which narrative elements made information less coherent or even more difficult to understand, for example, when the content is perceived as a kind of "advertisement" in which "the writing is invariably positive and sometimes kitschy" (okdkl52), or when "rambling descriptions are perceived as disturbing and artificial" (15nlvdq). The right balance between narrative elements and factual information in stories is clearly recognized by recipients by a "touch of reality", when the narratives seem realistic and meaningful to them from the perspective of their own lives. This touch of reality is created by the protagonists relating their experiences, in which information is not factually presented, but is "part of the story" making it "captivating" (tvjf5pv).

The results of the study show that storytelling is successful not only when it reflects one's own life through the protagonists' worlds of experience and meaning but when it can also be imagined as happening in one's own life. Such testing by recipients indicates a critical engagement with storytelling-based media calling for credibility and authentic representation (Gerrig & Prentice 1991). It becomes clear that narratives in the field of SusTelling not only have to meet the need for entertainment among their recipients but must also provide a demonstrable informative – and thus cognitive – use (Diffey 1995).

SusTelling Should Be Differentiated and Rich in Perspective

Another important aspect for participants is that the narrated content described should be as varied and differentiated as possible. With a contrasting presentation of perspectives or opinions, the participants want to get a comprehensive picture informing them about socio-ecological issues. They prefer "thought-provoking stories that give the reader a chance to get a different perspective on things" (8q5wyxq). This requires that a piece of information be differentiated, that is, "illuminated . . . from at least another side" (etyhry). Stories should take "different points of view" (5n4yama) and be told "without moralistic finger-wagging" (iwi6q43), in which "it is at least mentioned that this isn't now somehow a miracle cure for all problems" (etyhry). According to our findings, a diverse range of information that allows room for multifaceted interpretations and inferences is an important component of SusTelling. In contrast, narratives are evaluated negatively if they are "too superficial or a bit too . . . simplistic" or if they are "too exaggerated or too much is interpreted into things" (pa86o5m). Recipients find it disturbing when a topic or problem is not presented from several different perspectives. When "quite blatantly only one side is illuminated" (etyhry), recipients develop doubts about the credibility of a narrative, again bringing to the fore the importance of balancing narrative stylistic elements and factual information.

> That bothers me sometimes, when I realize, hmm, for this topic, I somehow have five things that immediately come to mind, which could have negative consequences and nothing is mentioned? Then I'm just naturally more skeptical about it.
>
> *(etyhry)*

The effect of multi-sided messages on media reception is addressed in the so-called Yale approach (Crowley & Hoyer 1994). It is argued that multi-sided messages are more likely to be persuasive and lead to changes in attitude among recipients than one-sided messages. The latter is more likely to create a feeling of being influenced. This finding can also be found in the results of the present study. Normativity and intentionality should indeed be central components in sustainability narratives such as SusTelling (see Chapter 1). However, the positive perception of such narrative content is reversed if it is too one-sided and contains normative representations of social and environmental problems. In this case, narrative messages do not give recipients the interpretive space they need and compromise credibility. Conversely, offering contrasting narratives increases the possibility of a reflective discussion among the participants. This is seen in various statements by participants emphasizing the beneficial effects of multiple perspectives: You can "better develop an opinion on the topic" by being "more stimulated to think" and "the topic sticks better in your mind" (9umi52r). In the encoding–decoding model that has become well known in reception research, this critical–reflexive type of argument is understood as a "negotiated interpretation" (Hall 1980, p. 136 f.), which is to be distinguished from a "dominant" and an "opposing" interpretation. This means that views and arguments in media content are examined against the background of one's convictions and views. Such reception behavior as an emancipatory process in that one consciously distances oneself from "hegemonic discourses" to develop one's stance is considered an important feature of critical media literacy (Kavoori 1999; Tisdell 2008). One function that the dramaturgy of a story can perform is to present information in different ways, weighing pros and cons and making complex relationships and problems visible (Grishakova & Poulaki 2019). For example, problems and risks (negative) in narratives can be contrasted with an alternative scenario (positive), which enables recipients to interpret them from different perspectives (Enschot et al. 2019).

Who Does SusTelling Reach and How?

So far, we have examined different characteristics of storytelling to determine which ones resonate with the participants, which provides clues as to what elements are important for creating successful sustainability narratives in the target group of young adults. It is now necessary to see how recipients process storytelling content. Three types of reception are found in the data (Figure 3.1), each of which shows a different approach to and use of SusTelling.

Explorers, visionaries, and *enlighteners* represent similarities and differences in the individual reception of sustainability narratives. The three types are ideal types that are not mutually exclusive. So, young adult recipients can never be assigned to just one type, since reception depends strongly on their interest in and prior

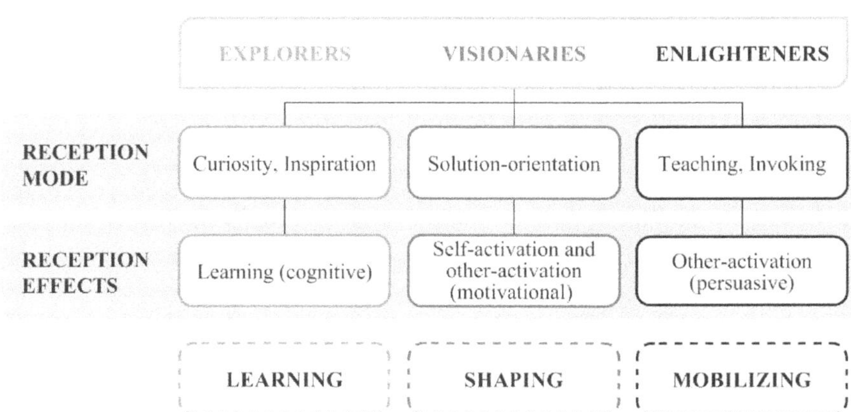

FIGURE 3.1 SusTelling reception typology

knowledge of a sustainable consumption topic and different expectations concerning reception. Preferences for certain media as well as situational factors (e.g., attention span, cognitive performance) – which could not be investigated within the scope of this study – also determine how participants engage with a text and how they were assigned to a type.

Explorers

For explorers, innovative information is an important criterion of SusTelling. Explorers are interested in the possibility of gaining new or greater knowledge about a topic or subject. Driven by curiosity, they are looking for inspiring insights in which "new points emerge in some form" (oefozve), or how to "apply" information to what they have "already learned" (oefozve). For example, one participant described, "And I always liked it when I read about things I didn't know before" (oefozve). In contrast, SusTelling is perceived negatively by explorers when a media article "didn't provide any new insights" and information is perceived as "irrelevant" (o0dheys). Narratives also serve for explorers as an impetus for new "food for thought" (79xpdas) "that were not yet as well known or popular in the general public" (pa86o5m). Stories in the field of sustainable living and consumption also serve as a source of consolidation or reinforcement of (newly acquired) views and beliefs among explorers. For example, one participant who has viewed media content offering a critical view on second-hand clothing reflects on her consumption behavior regarding the supposedly sustainable purchase of used items and arrives at a new perspective:

> This article is absolutely thought-provoking. Having already bought and sold on *Kleiderkreisel* [now: Vinted] myself, I now feel manipulated in some way. The company's marketing has apparently done a good job so that countless gals have a good feeling when they swap clothes via the app. The unacceptable conditions for seamstresses in faraway countries probably won't change. Greenwashing has once again worked.
>
> *(8q5wyxq)*

Visionaries

Visionaries respond to a constructive presentation of content that is designed to tell sustainability issues in a "solution-oriented" way (oefozve). One participant positively highlights such a narrative style in a storytelling text as follows:

> A shift in thinking needs to happen in many areas, and it's always nice to see that this is becoming more common and that more people are committed to it and are creating and living diverse solutions.
>
> *(ety8hry)*

Visionaries prefer narratives with a positive tone in which an orientation toward the future is expressed and the responsibility for sustainable ways of living is in focus. One participant states: "Super interesting post and a great idea that should also . . . be put into action" (8q5wyxq). Sustainability narratives further encourage recipients of this type to address ideas or needs for solutions concerning a specific problem area, for example, a participant who is critical of the view of mobility strategies offered in a visual narrative:

> It's a nice [hidden object] picture that depicts a sustainable utopia in the future. However, I think that cars don't have any place in a sustainable world. Innovative solutions need to be found here, as electric and hybrid cars also pollute the environment.
>
> *(8q5wyxq)*

Sustainability-oriented stories can also have a motivating or mobilizing function for visionaries when they present convincing stories about solving problems and visions of the future. As *unique* examples, SusTelling narratives create a view for *general* changes by inspiring sustainable ways of consumption and living or discussion:

> And I think it's cool when any ideas or tips are given about what you can do yourself or that you think over and say, oh, I could try that myself. So, I think that already shows that a text then had potential or achieved something.
>
> *(tvjf5pv)*

Storytelling articles are not only used by the interviewees to generate ideas for their sustainability actions. Insights into sustainability narratives are also used in discussions to influence others:

> And then you might also gather inspiration on how to talk about the topic better with other people. If somehow sentences or formulations or examples are mentioned . . . that would perhaps also convince this person in my life . . . [or] spend time thinking about it or something.
>
> *(ety8hry)*

Enlighteners

Participants who engage analytically with sustainability issues take on the role of *enlighteners* in the reception of SusTelling. They use sustainability narratives to call on their own often extensive and profound knowledge of the environment and sustainability and use this knowledge to point out existing problems and negative developments, which they reflect on in societal discourses (such as the role of digitalization, consequences of capitalism). These processes of reception reveal critical insights that are often combined with calls for changes leading to sustainable development. One participant expresses such a process with a text on sustainable fashion consumption as follows:

> Fair fashion is a super important topic, as the way we treat clothes is disrespectful and unsustainable. Due to digitalization and the fast pace of life in the world, we always have the desire to buy something new and the urge to keep up with the latest fashions. . . . Working together with mutual trust toward a goal to make a statement against capitalism and the self-centeredness of people is hugely important.
>
> *(8q5wyxq)*

Enlighteners see themselves as providing the impetus needed to motivate others or society as a whole: "It's time for humanity to realize its responsibility and address such problems more energetically together" (p58sgby). By taking a committed and idealistic stance on ecological problems and challenges, they draw attention to problematic situations. To this end, they appeal to (generalized) others to participate in the processes of change needed in many areas. Typical in the reception of *enlighteners* is a socio-critical engagement with the contents of storytelling items, reflecting on the absence or necessity of sustainable consumption, for example, through rhetorical questions: "Can a vacation be 'sustainable' if people travel by plane?" (8zck99q). Often, their reception is accompanied by an indignant attitude:

> Sad and shameful to see something like this. It's not just happening at the leather factory, so it's about time more was done!
>
> *(rrbam4k)*

In this way, *enlighteners* take a pedagogical approach, drawing attention to necessary changes and solutions to ecological problems:

> I think that the problem of unstoppable tourism can become a danger in many cities. More and more people are traveling without taking into account the people who live in the country or the environment. What is lacking is an awareness of travel and effective measures to prevent tourism from taking over locally. Tourists need to start developing empathy and understanding for local citizens and put themselves in their place instead of ceaselessly chasing the tourism craze regardless of the consequences.
>
> *(8q5wyxq)*

Explorers, Visionaries, and Enlighteners Compared

While for explorers learning and gaining new insights are at the center of the reception of SusTelling, for visionaries and *enlighteners* motivation is an important aspect in media reception – with a different orientation in each case. Enlighteners see themselves as providing ideas to mobilize others or society as a whole to more sustainable ways of living. Driven by their convictions, they call for a transformation of society and appeal to others or society as a whole. Visionaries, on the other hand, reflect more strongly on their actions and behavior. They expect SusTelling to provide them with innovative proposals for solutions to environmental problems that will help them shape sustainable developments. Inspired by successful SusTelling, visionaries want to rethink or change their behavior and enter into a dialog with others about these changes.

In addition to the insights into the potential cognitive (explorer), motivational (visionary), and persuasive (enlightener) impact of SusTelling, the type-dependent modes of reception also reveal how different competencies are applied. These are highly relevant for dealing with questions of sustainability (Wiek et al. 2011).[2] An anticipatory competency can be observed in particular in the case of visionaries, who focus on the solution- or future-oriented narrative content and classify it in terms of its practical feasibility. This requires a strategic competency to try out and implement innovative ideas obtained through reception. *Enlighteners*, on the other hand, have an educational perspective that brings normative competencies to bear by focusing their reception on collectively necessary processes of change and the need for action to shape sustainable development. In the cognitive acquisition of information and knowledge, explorers apply a systemic thinking competency in their engagement with SusTelling. Their focus is on gaining a better understanding of contexts or being stimulated by narratives to want to learn more or something new about a topic.

3.3 Conclusion

What conclusions can we now draw from the findings regarding the potential of SusTelling? The reception of SusTelling in the context of the study has shown that in addition to an authentic and subjectively experienced approach, a balance between narrative techniques and realistic information in stories is considered important by participants. Also, a narrative style that is as varied and differentiated as possible is considered an element of successful storytelling. This makes it clear that, in addition to the successful use of certain stylistic devices and narrative techniques, the informative content of narratives must stand up to critical scrutiny to arouse the interest of young people in the target group we studied. On the other hand, a comparison of the three different types of reception has shown that, in addition to individual learning processes, the (co-)shaping of socio-ecological change processes and the mobilization of others for this purpose represent important aspects in the reception of SusTelling. From this, it can be concluded that, from the point of view of the participants, storytelling

- arouses curiosity and stimulates learning processes (*explorers*),
- motivates people to implement new ideas and solutions to help shape sustainable development processes (*visionaries*), and
- creates stimuli to mobilize others or society for such processes (*enlighteners*).

The present findings thus provide insights into how to achieve accessibility to the target group of young adults and which focus narrative tools need to be able to influence new or existing views and competencies in favor of sustainable development.

A limitation of the findings relates to the homogeneous target group in terms of milieu affiliation and formal education level. The participants were largely members of educated middle-class milieus. And the findings clearly show that these are young adults who are already interested in topics related to ecology, the environment, and social justice and who are already actively engaged with sustainability issues in one way or another. This one-sided sampling results in a "thematic bias" of the participants (Graumann 1992), which influences not only topic choice but also the duration and intensity of their intrinsic attention to a media event. This phenomenon is also known as an echo chamber (Nguyen 2020) when media use consolidates or even strengthens existing beliefs and views in a closed circle of users who actively seek out information and media to confirm their beliefs and views. Our results also show this echo chamber effect as only people who already respond to environmental and sustainability issues and are active in this field were ultimately part of the study.

Given this limitation, we were not able to show (as we had initially intended) whether and to what extent different educational levels of young adults influence modes of perception and the impact of storytelling. For follow-up research, this raises the challenge of engaging audiences in this type of research who are not already predisposed to the topic to find out how less sensitized milieus access sustainability issues.

Notes

1 As a complement to the experimental study presented in Chapter 2, this exploratory study was intended to investigate the perception of SusTelling in the context of a diverse range of media and topics.

2 An approach to classifying competencies now widely used in sustainability research comes from Wiek et al. (2011). In addition to abilities to think through future scenarios and forecasts and deal with the uncertainties involved (anticipatory competency), an important ability is to be able to evaluate the current state of natural and social systems under the principles of environmental justice (normative competency). Complementing this, developing and trying out sustainable alternatives (strategic competency) and being able to communicate and collaborate with people on sustainability issues (interpersonal competency) are important competencies for addressing sustainability issues. This includes being able to understand the interrelationship of different spatial and temporal interactions across different social and natural systems (systems thinking competency).

References

BMU – Bundesministerium für Umwelt, Naturschutz und nukleare Sicherheit (2018). *Zukunft? Jugend fragen! Nachhaltigkeit, Politik, Engagement – Eine Studie zu Einstellungen und Alltag junger Menschen* [Future? Ask the Youth! Sustainability, Politics, Engagement – A Study on Attitudes and Everyday Life of Young People]. Berlin: BMU.

Bucher, H.-J. (2017). Understanding multimodal meaning making: Theories of multimodality in the light of reception studies. In O. Seizov & J. Wildfeuer (eds.), *New Studies*

in Multimodality: Conceptual and Methodological Elaborations (pp. 91–123). New York: Bloomsbury Academic.
Cohen, J. (2001). Defining identification: A theoretical look at the identification of audiences with media characters. *Mass Communication and Society, 4*(3), pp. 245–264.
Crowley, A. E., & Hoyer, W. D. (1994). An integrative framework for understanding two-side persuasion. *Journal of Consumer Research, 20*(4), pp. 561–574.
Dahlstrom, M. F. (2014). Using narratives and storytelling to communicate science with nonexpert audiences. *PNAS, 111*(4), pp. 13614–13620.
De Jager, A., Fogarty, A., Tewson, A., Lenette, C., & Boydell, K. M. (2017). Digital storytelling in research: A systematic review. *The Qualitative Report, 22*(10), pp. 2548–2582.
Diffey, T. J. (1995). What can we learn from art? *Australasian Journal of Philosophy 73*(2), pp. 204–211.
Enschot, R. van, Boogaard, I., Koenitz, H., & Roth, C. (2019). The potential of interactive digital narratives. Agency and multiple perspectives in last hijack interactive. In R. E. Cardona-Rivera, A. Sullivan, & R. M. Young (eds.), *Lecture Notes in Computer Science. Interactive Storytelling* (pp. 158–169). Cham: Springer International Publishing.
Esposito, E. (2007). *Die Fiktion der wahrscheinlichen Realität* [The Fiction of Probable Reality]. Frankfurt a. M.: Suhrkamp.
Eurostat (2020). *Being Young in Europe Today – Digital World: Eurostat Statistics Explained.* [https://bit.ly/3F7LAa5]. Accessed on May 20, 2022.
Fischer, D., Grunenberg, H., & Ruckelshauß, T. (2017). *Was wollen Kinder und Jugendliche in Bezug auf Nachhaltigkeit und Konsum wissen, wie kommunizieren sie und wie informieren sie sich? Eine Literaturanalyse* [What Do Children and Young People Want to Know About Sustainability and Consumption, How Do They Communicate and How Do They Inform Themselves? A Literature Analysis], INFU Diskussionsbeitrag 40 / 2017. Lüneburg: Leuphana Universität Lüneburg.
Frey, F. (2014). Wirkungen des Narrativen. Ein systematischer Forschungsüberblick zu Effekten narrativer Kommunikationsformen [Effects of narrative. A systematic research overview of effects of narrative forms of communication.]. In W. Früh & F. Frey (eds.), *Narration und Storytelling. Theorie und empirische Befunde* [Narration and Storytelling. Theory and Empirical Evidence] (pp. 120–192). Cologne: Herbert von Halem Verlag.
Gerrig, R. J., & Prentice, D. A. (1991). The representation of fictional information. *Psychological Science, 2*(5), pp. 336–340.
Gossen, M., Holzhauer, B., Schipperges, M., & Scholl, G. (2016). *Umweltbewusstsein in Deutschland 2014 – Vertiefungsstudie: Umweltbewusstsein und Umweltverhalten junger Menschen* [Environmental Awareness in Germany 2014 – In-depth Study: Environmental Awareness and Environmental Behavior of Young People]. UBA-Texte 77 / 2015. Dessau-Roßlau: Umweltbundesamt.
Graumann, C. F. (1992). Speaking and understanding from viewpoints: Studies in perspectivity. In G. R. Semin & K. Fiedler (eds.), *Language, Interaction and Social Cognition* (pp. 237–255). London: Sage.
Grishakova, M., & Poulaki, M. (eds.) (2019). *Narrative Complexity: Cognition, Embodiment, Evolution.* Lincoln: University of Nebraska Press.
Hall, S. (1980). Encoding/decoding. In S. Hall & D. Hobson (eds.), *Culture, Media, Language. Working Papers in Cultural Studies* (pp. 128–138). London: Hutchinson.
Hodson, J., Dale, A., Jost, F., & Clifton-Ross, J. (2020). Sustainability issue communication and student social media engagement: Recommendations for climate communicators. *Journal of Digital & Social Media Marketing, 8*(3), pp. 277–290.
Kavoori, A. P. (1999). Discursive texts, reflexive audiences: Global trends in television news texts and audience reception. *Journal of Broadcasting and Electronic Media, 43*(3), pp. 386–398.

Kolodzy, J. (2013). *Practicing Convergence Journalism: An Introduction to Cross-media Storytelling*. London: Routledge.
Krcmar, M., & Strizhakova, Y. (2009). Uses and gratifications as media choice. In T. Hartmann (ed.), *Media Choice: A Theoretical and Empirical Overview* (pp. 53–69). New York: Routledge.
Kress, D. (2021). *Greenpeace Sustainability Barometer 2021 – We are ready and finally want a sustainable future!* Summary. [https://bit.ly/3Agns33]. Accessed on August 22, 2022.
Kuckartz, U. (2014). *Qualitative Text Analysis: A Guide to Methods, Practice & Using Software*. London: Sage.
Levy, M. R., & Windahl, S. (1985). The concept of audience activity. In K. E. Rosengren, L. A. Wenner, & P. Palmgreen (eds.), *Media Gratifications Research. Current Perspectives* (pp. 109–122). London: Sage.
Lugmayr, A., Sutinen, E., Suhonen, J., Sedano, C. I., Hlavacs, H., & Montero, C. S. (2016). Serious storytelling – a first definition and review. *Multimedia Tools and Applications*, *76*(14), pp. 15707–15733.
Michelsen, G., Grunenberg, H., Mader, C., & Barth, M. (2016). *Greenpeace Nachhaltigkeitsbarometer – Nachhaltigkeit bewegt die jüngere Generation* [Greenpeace Sustainability Barometer – Sustainability Moves the Younger Generation]. Bad Homburg: VAS Verlag für Akademische Schriften.
Moyer-Gusé, E. (2008). Toward a theory of entertainment persuasion: Explaining the persuasive effects of entertainment-education messages. *Communication Theory*, *18*(3), pp. 407–425.
Nguyen, C. T. (2020). Echo chambers and epistemic bubbles. *Episteme*, *17*(2), pp. 141–161.
Reisch, L. A., & Bietz, S. (2011). Communicating sustainable consumption. In J. Godemann & G. Michelsen (eds.), *Sustainability Communication* (pp. 141–150). Dordrecht: Springer.
Schreier, M. (2012). *Qualitative Content Analysis in Practice*. Los Angeles: Sage.
Stevenson, K., & Peterson, N. (2016). Motivating action through fostering climate change hope and concern and avoiding despair among adolescents. *Sustainability*, *8*(1), 6.
Tisdell, E.J. (2008). Critical media literacy and transformative learning. *Journal of Transformative Education*, *6*(1), pp. 48–67.
Weckroth, M., & Ala-Mantila, S. (2022). Socioeconomic geography of climate change views in Europe. *Global Environmental Change – Human and Policy Dimensions*, *72*, 102453.
Wiek, A., Withycombe, L., & Redman, C. L. (2011). Key competencies in sustainability: A reference framework for academic program development. *Sustainability Science*, *6*(2), pp. 203–218.
Zillmann, D. (1996). The psychology of suspense in dramatic exposition. In P. Vorderer, H.-J. Wulff, & M. Friedrichs (eds.), *Suspense. Conceptualizations, Theoretical Analyses and Empirical Explorations* (pp. 199–231). Hillsdale, NJ: Lawrence Erlbaum Associates.

PART II

Practice Perspectives

4

TELLING YOUR OWN STORIES

Teaching and Learning Storytelling

Hanna Selm, Daniel Fischer, Heike Janßen, Torsten Schäfer, Anna Sundermann, and Sonja Fücker

4.1 Introduction

The first three chapters of this book have shown that targeted use of storytelling for sustainability (SusTelling) requires a multidimensional understanding that includes a content level, a value level, and a craft level. But how can learners be engaged in exploring these levels and their interplay? What experiences and approaches can be built upon to teach SusTelling in education and training in different fields of practice? This chapter explores these questions in two parts.

The first part consists of two stories, both of which describe challenges in teaching and learning storytelling from different perspectives. The first story (4.2) is from the perspective of a researcher and tells of an arduous and doubt-filled journey of developing an online learning unit to teach SusTelling to practitioners. In the second story (4.3), a journalist tells of their daily struggle to communicate the importance and urgency of sustainability issues in newsrooms, while facing very different challenges in teaching journalism students.

The second part of this chapter is a transcribed conversation in which the participants in the two stories exchange their understandings, challenges, dangers, and recommendations in communicating SusTelling. They return to aspects from the previous stories and provide greater depth. This chapter concludes with insights into what can be learned from the stories and the conversation about teaching and learning SusTelling.

4.2 How Can SusTelling Be Used in an Online Learning Unit for Practitioners: The Story of a Researcher

By Hanna Selm, Daniel Fischer, Anna Sundermann, and Sonja Fücker

DOI: 10.4324/9781003326144-8

> The SusTelling online learning unit is available free of charge as an Open Educational Resource in German and English at www.elearning.sustelling.de/en/. The learning unit aims to develop competencies for analyzing and producing sustainability-related storytelling narratives. The target group is people who want to learn how to communicate sustainability topics excitingly in narrative form but who have little or no knowledge of story writing. The learning unit consists of three modules, with each module taking about 15 minutes to complete.

Enter. It's loading. There it is, the SusTelling online learning unit. It's done. I'm satisfied. Again, some test clicks, as so often before. The layout is clear and appealing. I like the backstory as well as the structure. All these examples make it interesting. My view wanders out of the window over the campus, which shines in spring-like colors. Down in the campus garden, the pretty beehive boxes are back. As always, as I watch how busy the bees buzz about. I'm fascinated by how unperturbed they go about their work, collectively playing a crucial role in the ecosystem that surrounds them. I'm ashamed of how humans intrude on their habitat and show little interest or gratitude for their beauty and what they do. Not only that, but I pause and feel how quiet it has become around me and within me. I think back on what a long way it was to get here. A journey into a world that was foreign to me. These thoughts make me smile.

It started in the fall of 2016 when we received approval from the funding agency. We had set our sights high. Not only did we want to do research, but we also wanted to share what we would find out with people in the field who would be able to make use of our findings. At least, that's what we hoped. None of us had the slightest doubt about the importance of what we were embarking on.

Looking back, the first half of the project flew by. We were busy systematically evaluating the available research literature, developing our empirical studies, and discussing our findings. We turned to the question of how we should communicate what we had discovered to the people involved in communicating sustainability issues daily – whether in companies, the media, or academia. One important product we had in mind was an online learning unit. We wanted to give all those who don't have the time or money to develop competencies in sustainability communication a compact, free tool. One that was practical and could teach them how to communicate their projects and initiatives more excitingly and engagingly. Anyone and everyone – that was our idea – should be able to use this learning unit to learn something about what SusTelling is and how it can be used at their own pace, anytime and anywhere.

When we started to work more intensively on the online learning unit, I realized that our learning unit would be a big milestone not only for the project but also for me. The goal was clear, but what was to be learned and taught, and how? Two concerns were paramount. First, the learning unit should help learners understand how to use SusTelling and what its effects are, and, second, they should be able to practice writing stories themselves. Honestly, I have to say that I struggled with the second goal. I absolutely agreed that it was essential for users to learn how to use it practically. And yet, teaching writing skills was not our craft, not our expertise. At our weekly team meeting, we wrestled for a long time about what exactly the goal of the online learning unit should be, until we finally found a compromise that took both analytical and creative skills into account. Knowledge transfer means making compromises and – at least to some extent – leaving one's academic comfort zone. More challenging than the question of what should be learned, it later turned out, was how it should be learned. It was about pedagogy.

We already suspected – and know much better today – that there are differences between teaching in-person seminars at the university and in a learning unit online, between students and a largely unknown target group, and between teaching and learning together in a classroom, online, and self-taught. What exactly these differences were and how they should influence what our product needed to look like was not really clear to us at the beginning. So, we turned to professionals. Fortunately, we had the chance to get Denise, an e-learning expert, and Kai, a programmer, on board to implement the learning unit. What followed was a steep learning curve for all of us.

It was that one virtual team meeting when Denise caught us unawares. She kindly but firmly pointed out our own contradiction in the design of the learning unit. That we planned a learning unit about storytelling in which we did not use storytelling ourselves. We were defenseless. I remember well how, even during the conversation, counter-arguments were quietly gathering in me: "To simply pack all of the results of the study into a story wouldn't do justice to its complexity. It would mean narrowing what we want to communicate to a story and a specific audience – this would undermine our understanding that doing science involves making cross-contextual claims. A story also means using language in a loose way. It wouldn't be precise enough!" In short, at this point and with my self-understanding as an academic researcher, I didn't warm up at all to the idea that storytelling was the appropriate way to communicate our project findings. It was unsatisfying. I closed my notebook, drove home, and tried to gain some perspective.

As so often happens, after the confusion things became clearer inside me. The next sunny morning, I sat in my office with the beautiful view I still can't get enough of. I thought of Denise, of the contradiction she had pointed out to us, and how we were resisting finding a solution. It took some time, but finally, she had convinced me. So, we shortened, changed, and redesigned the unit. The result was a backstory written by us. That's what we called our SusTelling-style story, in which the protagonist Anna wants to open a package-free store and learns SusTelling, so she can present her initiative appealingly and authentically in the local newspaper.

I saw the light. It was wonderful how this story connected all three modules making up the learning unit. I also liked the fact that the backstory showed how what we wanted to teach could be used in real-life practice. And, finally, with Anna, we now had a person we hoped the users could identify with. I thought I now really understood Denise's insight.

A bright reflection in the campus garden caught my attention. How could we prevent our storytelling learning unit from being misused as a tool to subtly influence people? Then it became more than clear to me that the learning unit had to create a reflexive attitude by confronting learners with questions like, why do I want to use SusTelling? How do I understand sustainable development? What impact do I want to achieve?

The protagonist of our backstory – that's Anna – offered what I thought was a teachable moment when she could give her answers to these questions and show what they meant in her situation. It felt right to build these opportunities into the backstory. No finger-wagging, no lecturing, but an "invitation" to explore the SusTelling learning path for themselves. Learners should be able to decide for themselves how fast, how deep, and in which order they progressed through the unit.

Deeper and deeper we dived into a kind of teaching that was new to us – and it was and still is a joy. It feels like a real transfer, from science to practice, but also from e-learning practice to science. I'm still sitting in my office, looking out at the greenery, and I'm smiling to myself. This experience was like a small whirlwind that, at first, I wanted to seek shelter from. Now I'm glad we got carried away; it was inspiring.

Above all, I took away some basic questions that I resolved to ask in future SusTelling teaching contexts. What are my learning goals? How should I design the content, so the audience will connect with it? How can I create authenticity in my teaching? What assumptions am I making about learners? Where and how do I relate content to sustainable development? And, finally, how can I create a learning environment that works? I have learned to ask these questions again and again, and not to simply take my past teaching-learning experiences and apply them to new and different learning contexts.

My gaze wanders to the screen. Enter. The online SusTelling unit is loading. Now I understand it is one possible answer to these questions, and I'm curious to see whether it finds its place in the world out there, meets with resonance, and perhaps can even help one initiative or project to make how they communicate sustainability more engaging and lively. Again, the reflected sunlight from the campus garden flashes into my office. I take one last look at the homepage of the learning unit, close my notebook contentedly, and decide to buy some groceries on the way home. Package-free, of course.

4.3 We Are Sisyphus: A Story of the Difficult Attempt to Spread the Word About Sustainability Journalism

By Heike Janßen and Torsten Schäfer

> Heike Janßen and Torsten Schäfer are founding members of Netzwerk Weitblick-Verband Journalismus & Nachhaltigkeit (short: Weitblick Network), the German association for sustainability journalism. As part of the association's training efforts, they wrote a book (in German) that provides a comprehensive, yet low-threshold introduction to journalistic storytelling on sustainability issues. The book is a toolbox that offers a great deal of practical help and advice in 17 chapters on how sustainability-related storytelling can be learned and communicated.

How do you tell a story about writing your own book about storytelling in a book that is also about storytelling? Of course, we first looked for the elements of a good story, the tools of storytelling: we, the authors, are the "heroes" of the story. The plot is about the process of writing. And the dramaturgy, the resistance that exists in every good story? It would be our laziness, the blank screen we sit in front of until, sooner or later, an idea comes along.

Kind of boring, isn't it? This is how everybody does it. It's not new, no new insights, no new perspectives, no real conflict, no really interesting ideas. So, what's the goal, the dream, the quest of the hero, the author? What is it that drives them? What is the deep "need" behind the goal, behind the "want". So the "want" is the goal the hero sets themself: defeating their opponent in the boxing ring. But the "need" is deeper than the "want." It's the love the boxer has for the mousy-looking saleslady in the pet store next to the boxing gym. We, the viewers, know that the "need" is the important thing, not the "want." We, as authors, realize it was not the "want" but the "need" that drove us to finish the book. It's the "need" that drives us, the need to try to save the world. For our children. That's what's at stake, the future of our loved ones. Our contribution is what we can do as teachers. We can change how journalists report the news.

The equation that we need to solve is: journalism + sustainability = more information, more attention, more understanding = potential change → saving the world.

It's heroes like Bruce Willis that save the world. Though Greta is a hero, a real one too. She's taking on and wrestling with enormous problems and everything is at stake. Climate catastrophe, species extinction, and other ecological crises – humanity has put itself in an unprecedented situation – comparable to sailing the Titanic toward an iceberg. But there are no real superheroes like Bruce to save us. We have to become heroes, just like Greta and the many other everyday heroes. We have to save ourselves.

Everyone has a part to play. Each person has their abilities. As media people, we, Torsten and Heike, can help shape public opinion with facts and information. By explaining to people that things can't go on like this. Because if we all continue as if nothing has changed, the world as we know it will end. Climate crisis, species extinction, it's already here today. In other regions of the world, it's already much worse, people, animals, and plants are dying. It's not a movie. It's our reality.

But how do we explain it so that people will not only read about it but internalize this message and be motivated to carry it to others? Or even change their behavior? This is the question that guides our reporting, and also the seminars and courses in which we pass on our knowledge of how to be a journalist.

There is one major obstacle to this approach. Many of our colleagues, including the younger trainees, have not yet really understood that the connection between the climate crisis, resource depletion, and species extinction is the most important overarching issue of our time. The framework in which we must scrutinize the economy, society, and politics. And, most importantly, sustainability is not about a single issue, it's a big environmental and social dimension, a context in which all other issues are played out. The Guardian newspaper has recognized this: We should report on it the way media did before the outbreak of World War II. Every day it should be the lead story. Not the stock market report before the daily news, but the state of the earth and its inhabitants. But the reality in the editorial offices is different. The climate crisis, resources, and migration – these are all topics among many. On a par with sports, the news, and entertainment.

Climate topics are actually sexy because they have a lot to do with lobbying, corruption, and cronyism. Still, when you tell investigative journalists about it, they get a glazed look and think of something more "important" to do. In our profession, we climate people are still the annoying ecologists constantly trying to give others a guilty conscience. Sure, they say, it's all super important. But most of the time they live a different life. Weekend trip to London or Barcelona? Sure! Giving up eating meat? Oh no!

That's our challenge.

In 2015 we founded Netzwerk Weitblick (www.netzwerk-weitblick.org), a non-profit network for sustainability and journalism, together with about 60 other journalists. We want to inform and educate our colleagues on sustainability issues. Because anyone who knows and has really felt, who has recognized what is happening to the planet and humanity right now – and knows what needs to happen now – has no choice but to report differently, and see things differently. But it is still hard for us to find sponsors and educational partners. Few understand that knowledge is needed to do a good job reporting on sustainability. That the PR people hired by the coal industry or the SUV builders and their lobbyists do a good job too. It's demoralizing that the narrative is still that the energy transition is expensive. That in a free country, drivers should have the freedom to decide how fast they drive. And that far too many journalists are generalists who aren't always able to weigh facts correctly. Not because they're not smart enough, but because they lack the research, specialization, and advanced training in areas like sustainability and storytelling. Because it's not worth it. Because they don't have the time.

But slowly there are more of us. In seminars, we write scenarios for climate justice. You could say, we write stories about a desirable future. Writing scenarios gives us courage and is appreciated by our readers, but there has to be more than this because our editors want reports and commentaries. But imagining a different world is the basis for other forms and other ways of working. So is writing in nature, for instance, where students come to write in a very different, more vivid language. One that is more accurate and more beautiful than officialese or scientific jargon – everything that ruins a good story. But then we use a mix of methods to find everything that works in writing. We have a debate, a fictive editorial conference, train using synonyms or finding the perfect phrase, writing bad texts, or doing pair editing.

We also work in a very structured and focused way. With public relations people from institutions and NGOs, we use the tools of storytelling in intensive courses. We track down heroes, develop new perspectives, and find the best dramaturgy and the core message. We find good images for photo spreads and videos and develop new social media content. We write good stories to stand out from the background noise of information.

But we, Torsten and Heike, are not heroes. We are Sisyphus. And our story will probably never end. Because we keep going in the hope that at some point we will have a greater impact on our profession. We are encouraged by small successes, the creation of the Weitblick Network, more and more seminars, media awards, start-ups, and online media – all of which have emerged in the field of sustainability over the past ten years. We are also encouraged when structures in the media and at universities change – whether in Lüneburg or Darmstadt – and become more sustainable.

But then the old arguments reappear and old doubts arise even about the very sensible idea of presenting complex topics better – of telling stories. Just because a reporter made up a lot of stories, some people now doubt the integrity of storytelling itself – and so doubt how we as journalists like to write and as instructors like to teach about sustainability. Because good stories communicate information better, they're more gripping and have more impact, and they encourage people to reflect and take action. It's always about the information we're communicating. Being entertaining and telling a good story always comes second. It takes work to explain something well. But we try. And when sometimes we fail we try again. Perhaps that's what our story about telling stories is about. It's a never-ending story about a story that's badly needed.

4.4 Teaching and Learning SusTelling: A Conversation

Why Did You Tell the Story the Way You did?

Heike Janßen and Torsten Schäfer: "We are convinced that journalism about sustainability is incredibly important and that we are in the process of constantly writing this story about the development of sustainable journalism. From a craft perspective, it wasn't easy to write this story, as there are no classic heroes in sustainability. Mind you, we see ourselves as journalists, not heroes. We've tried to show journalism in the sustainability field as a lonely hero and a Sisyphus. Because we wanted to highlight the strong headwinds we face. But now and again Sisyphus can also celebrate successes, such as the emergence of the Weitblick Network, which provides a tailwind. As with Sisyphus, however, the question of the larger meaning arises. The story should make clear why journalists continue despite opposition and setbacks, and what drives them to report on sustainable development. For us, there is this deep need to stand up for a world in which everyone wants to live".

SusTelling project team: "In their story, Torsten and Heike graphically described the external struggles that are inevitable if we want to give sustainability a greater weight in communication so it's heard by more people. For us, it was clear early on that we wanted to dedicate our story more to the inner struggles of teaching and learning SusTelling. What goals should we keep in mind when we teach SusTelling? How can we authentically implement what we teach? And how can we deal with the normativity that constitutes SusTelling? We wanted to write about the teaching challenges we wrestled with as we developed the online SusTelling unit. At the core of the story, we faced three challenges: embedding the content in a backstory, teaching a reflective attitude, and developing an understanding of learning. The story is meant to trace and relive how we arrived at what is now seen as a product".

What's Special About SusTelling Versus Storytelling? Is There Something Special About It?

Heike Janßen and Torsten Schäfer: "In contrast to storytelling in general, SusTelling has a normative, ethical superstructure with the core values of sustainability. These core values are the 'backyard' or the 'heavy load' that are always present in every story. A story that is limited to the decorative flowers in the front yard doesn't work with SusTelling. And SusTelling has an educational goal. On the one hand, it's important to make people understand the alarming state of our world, but it's equally important to motivate them positively. Creating a sense of community and pointing out possibilities for action form are crucial to this motivation. We need to show that we're not alone in bearing the weight of sustainability, nor are we powerless or helpless. To convey these central elements of sustainable storytelling based on scientific facts, and to do so without creating the illusion that everything can be solved very easily, is the tightrope act, the art, the challenge of SusTelling. In this respect, the question of where we're going is a genuine element of SusTelling, which it isn't in storytelling. It also becomes clear that SusTelling is always a movement from the periphery to the center, from the minority to the majority, whereas storytelling doesn't have to move this way".

SusTelling project team: "What Heike and Torsten said about the 'backyard' and the 'educational' in SusTelling reflects very well what matters to us. SusTelling wants to make a double contribution by supporting social change toward sustainable development and always aims at making an educational impact as well. For SusTelling it's not enough to subtly influence the behavior of the audience – for example, by making a certain product appealing. In SusTelling we always aim for reflexivity as well. This is why we have also placed this educational and sustainability effect at the center of our definition of SusTelling [Chapter 1]".

What Are the Challenges of Teaching and Learning SusTelling?

Heike Janßen and Torsten Schäfer: "In teaching SusTelling it's crucial to always involve a perspective of the future, not as something miserable, but as a desired state – What do we want the world to look like in 2050? It also seems time to remake the classical hero figure in SusTelling – something that also needs to be part of teaching. This is about replacing the shining hero (president, queen, world boxing champion, explorer, and so on) in favor of a hero figure that implies participation, community building, and the collective spirit. The typical SusTelling hero shines through reaching goals together, being humble, and being free".

SusTelling project team: "We think that three challenges should be considered in teaching and learning SusTelling. First, SusTelling requires a high level of expertise on sustainable development issues. Second, a reflective attitude, in the sense of having a normative goal, is essential. This includes dealing with SusTelling's undesired effects and 'collateral damage'. Third, SusTelling is always a balancing act between the imperatives of objectivity and the limits of the storytelling technique, between creativity and craft".

Storytelling Is Accused of Being a Social Technique to Influence People. Where Do You Stand on This?

Heike Janßen and Torsten Schäfer: "Storytelling is an age-old social and cultural technique that has been used in every culture to convey knowledge as well as norms.

However, it can also be used free of interpretative, implied, or normative contexts if one just makes use of its elements, such as knowing the characteristics of suitable protagonists, the appropriate dramaturgy, and finding a challenge with something at stake, or identifying the story's turning points. SusTelling, on the other hand, has a normative context given by sustainable development that serves as a value and an educational goal.

There are different views in the media scene. Should journalists simply depict reality objectively and have personal values only as a private person? That's the approach in the United States, among others. At the extreme, this has led to the arguments made by climate change deniers being presented on an equal footing with the overwhelming majority consensus in science. One thing is clear. Journalism should enable people to form an opinion based on objective reporting. Nevertheless, journalists must make a sensible decision about whose contribution to the topic is meaningful and qualified. Journalism also always operates within a social and cultural framework based on fundamental values. Here in Germany, we operate within the constitutional rights of a free democracy, so most follow a recognized code of values on issues such as racism and also sustainability issues (human rights, right to clean water, right to education, and more). These fundamental rights and values manifest themselves on, among others, a legal, social, and political level. Many colleagues also see it as the task of journalism to communicate and protect these values. It's important to note that this is a matter of values, not opinion. But how this value is expressed remains the responsibility of the individual journalist and their editors.

And finally, the use of storytelling in journalism also harbors dangers. For example, the 'Relotius affair' [a journalist falsified magazine and newspaper articles, often using fabricated storified elements, on a grand scale for a major news magazine] teaches us that we must never smooth over the rough edges out of love for our story and contrary to the facts. This closely reflects our own story above because it represents the ups and downs of reality. The Sisyphus comparison we use in this story isn't exactly commonplace, as it doesn't trace the classic success story we would also prefer to tell. An explicit danger in SusTelling is that we filter what fits the story and our goal, and what doesn't.

In the field of sustainability, stories about some topics can be told well, while other important topics, such as land use, get too little attention because they're very factual and are less amenable to illustration. As mentioned earlier, there's also a danger of not integrating issues at all or integrating them inadequately into the larger context of sustainable development. To stay with the previous image, the backyard (the big picture) remains underdeveloped and the presentation is limited just to (front yard) stories that address individual issues but fail to frame them in the larger context. While such stories can appeal to audiences, they can also distort our understanding by losing the context of the bigger picture, that is, the broader sustainability challenge".

SusTelling project team: "In principle SusTelling as a tool is not immune to being used for uncritical, complexity-reducing, or manipulative purposes. We have built a 'safety mechanism' into our SusTelling definition by including reflexivity and reasoned evidence (as distinct from the blunt appeal to emotion) as defining characteristics. The trick is to compose stories that balance the content level (facts, context, rationale, and so on) and the story level (characters, plot, and so on) in ways that we are always able to justify why the story makes the audience more reflexive and the world more sustainable".

What Advice Would You Give to People Who Want to Learn and Teach SusTelling?

Heike Janßen and Torsten Schäfer: "In our practice, it's turned out to be very inspiring to develop stories collectively in groups. After all, stories have always been a social phenomenon that brought people together, transmitted knowledge among people and across generations, and helped align perceptions of the world. We mainly use shared creativity processes in groups for this purpose. If there is a positive, appreciative atmosphere, the group can help each other expand their imagination. One way to condense a story to its essence is to tell it to others over and over again so that it becomes clear what the core of the story is and where there are still gaps in the narrative or on the factual level.

In university teaching-learning contexts, we would recommend insisting that each level is considered. When building SusTelling stories, it's immensely important to always include references to sustainability and facts, then of course to work on the craft level, and last but not least to show concrete possibilities for taking action. These levels have to come together in good stories, and that can be learned".

SusTelling project team: "We would add that it's important to have a deep understanding of the audience and their needs. We all tend to become ego-centric and get stuck in our frames of what the problems, causes, and solutions are. The potential of stories, however, is precisely their complexity and immediacy in this respect – people can be stimulated by stories in very different ways and find very different messages and lessons in them. In a strategic use of SusTelling, it's enormously important to understand who we are writing this story for. Another important point, and one not mentioned so far, is authenticity. Real experiences and contradictions are particularly suitable for SusTelling. In other words, stories that contain a moment really worth telling. At this point, we would like to point out that not all experiences are suitable for a story. Finally, we would also like to encourage people not to limit themselves to telling success stories or stories where everything works out. Storytelling has a long and powerful tradition of expressing and denouncing injustice, whether injustice we have experienced in the past or ongoing injustice. In the future we would like to see more stories that focus on the structures that push and force us into collective 'unsustainability.' SusTelling in this regard can be much more than a way of educating others or spreading solutions; it can be an expression of collective suffering and rebellion against a world that is locking us into unsustainable practices and blocking sustainable change".

4.5 Conclusion: SusTelling Demands Competencies on Three Levels

What insights can be gained from these two practical examples about teaching and learning SusTelling? Both in the stories and in the conversation, it has become clear that successful SusTelling requires competencies on different levels.

- On the content level, factual knowledge about sustainable development issues so that they can be critically evaluated and linked within the story's framework.
- On the value level, a critical awareness of the manipulative risks inherent in narrative communication and a commitment to promoting sustainability in a reflexive manner.
- On the craft level, mastery of writing technique and expertise in constructing stories, as well as creativity.

The stories describe two very specific learning opportunities – the SusTelling online learning unit and the educational offerings of the Weitblick Network – that provide an entry point into learning SusTelling. However, the conversation also makes clear that the challenges in teaching SusTelling vary greatly depending on the context. This can be explained by how understanding at the content level, the value level, and the craft level is shaped by existing knowledge and prior experience. For example, the second story highlights the content-level challenge of communicating the meaningfulness and complexity of sustainability issues, while the first story focuses on the value-level challenge of guiding teachers in developing a reflective approach. The distinctions between teaching–learning objectives and between the concrete approaches proposed in this chapter can be used as an aid to better understand a target audience and to design teaching and learning processes so that they meet both the specifics of the target audience and the requirements of sustainability-related storytelling.

5

STORYTELLING FOR SUSTAINABILITY IN HIGHER EDUCATION

Petra Molthan-Hill and Daniel Fischer

Although teaching and learning have been taking place in higher education institutions for hundreds of years, ideas about what and how to teach and learn are constantly changing. This has to do with changing societal expectations about what students should know and be able to do after graduation. But it also has to do with advancing specialization in the sciences themselves, which are producing new knowledge at an ever-faster pace in often highly differentiated fields, while also developing new approaches and procedures to produce this knowledge. To cope with this complexity, storytelling has been receiving increased attention in higher education. In what follows, we argue that storytelling has a potential for teaching and learning about and for sustainability in higher education institutions that has only begun to be recognized.

We propose three approaches to using storytelling as a method for teaching and learning sustainability in higher education. First, storytelling can be used in a variety of ways to complement the teaching of factual knowledge. For example, learners gain a more holistic understanding of the problems related to child labor in chocolate production when facts are communicated using or supplemented by stories from a child's perspective (Manteaw 2020). Second, as a method for gaining knowledge, storytelling can also be used to open up new perspectives on phenomena and to document experiential knowledge, or lived experience. Third, storytelling is also a method for communicating with others. We often underestimate how strongly public opinion is influenced by narratives. This could also be taught at university – in all humanities, social and natural sciences – so that students can not only reflect on their own stories but also learn how storytelling or even Storytelling for Sustainability (SusTelling) can be used to communicate at the science–society–policy nexus.

The three approaches, presented in more detail later, also reflect the tension inherent in using storytelling for sustainability (see Chapter 1). Is SusTelling used to understand or change how people think and act about sustainability? And should

DOI: 10.4324/9781003326144-9

it be about learning and empowering individuals (as a pedagogical approach), or about empowering students to engage with audiences outside academia through the use of narrative communication (external science communication)? As university teachers with many years of experience in teaching sustainability topics at various international universities, we always see the use of storytelling for sustainability in a specially designed educational context. Part of this design is the orientation toward the normative idea of sustainability and its postulates of justice. In the same way, however, reflexivity is part of the pedagogical use of SusTelling. This means that it is always about understanding sustainability as a "collective process of searching, learning, and creating sustainable solutions" (Stoltenberg & Burandt 2014, as translated by Singer-Brodowski 2018, p. 1300), to which SusTelling contributes by helping to interpret, explain, and understand.

In this chapter, we want to highlight the variety of ways storytelling can be used in higher education institutions to promote sustainability and illustrate it with examples. We distinguish three ways of using SusTelling: as a method for teaching and learning (5.1), for gaining knowledge (5.2), and for communicating with others (5.3). In the conclusion, we advocate exploring the potential of the many different varieties of storytelling more proactively and strategically.

5.1 Storytelling in Higher Education

Silence falls in Dunedin, an old university town in southeastern New Zealand. The lecturer gathers together his students attending a course in clinical dental anatomy. He tells of what happened to him one Friday afternoon many years ago in South Africa, where he was then a dentist. Johnnie, a 5-year-old boy, was brought to the dentist's office by his aunt, who was agitated. The lecturer had treated Johnnie just a few days earlier and had given him a filling in one of his molars. Now Johnnie's entire face was swollen, and his lower eyelid was nearly closed from the severe swelling. At the sight of this, the lecturer noticed how nervous he suddenly became – was this an infection, perhaps even something serious like an infection of the submaxillary fossa (a shallow depression below the lower jaw that contains, among other things, the muscles used in chewing)? After some coaxing, Johnnie opened his mouth so the lecturer could get his mirror in place. He was relieved to see that there were no signs of infection and that there was nothing wrong with the tooth. He then spoke calmly to the boy, who, after some coaxing, finally told him that he had taken a honeycomb from a wild beehive and had been stung. He had been embarrassed, however, which is why he didn't tell his aunt, who then panicked and took him directly to the dentist's office. What the lecturer learned then was twofold: first, the value of a good medical case history and, second, that the swelling of an eyelid not only can be symptomatic of a submaxillary fossa if it is postseptal but can also simply be the result of a bee sting if it is preseptal. Students nodded and began taking quick notes as the tower clock on the main campus of New Zealand's oldest university struck, ending classes for the day.

The Dunedin story is featured in a journal paper in which a group of faculty shares their experiences using storytelling to educate medical students (Kieser et al. 2008). The example illustrates how stories and storytelling can help embed supposedly "dry" or "abstract" content in context so that it becomes more interesting and easier to remember (Glonek & King 2014). It is primarily this teaching potential that has brought some attention in recent years to storytelling in higher education.

Stories and storytelling are being increasingly used to communicate about issues in sustainability in higher education. The most recent and arguably most comprehensive collection of these examples is the edited volume *Storytelling for Sustainability in Higher Education: An Educator's Handbook*, published by Routledge in 2020 (Molthan-Hill et al. 2020). This volume includes both research findings and practical applications of storytelling. In the first part, the definition of storytelling as SusTelling is given a separate chapter, before approaches are discussed that show how current sustainability problems can be presented in the form of historical narrative or science fiction. Research is then summarized on the effect of utopian or dystopian narratives on learner behavior related to sustainability. The second part focuses on the direct use of stories in the classroom, for example, as the subject of an exam. In the third part, examples are given of how storytelling can be used in different disciplines, with a focus on those less likely to use narrative elements, such as accounting, chemistry, engineering, and law.

A prominent and trending variety of storytelling in higher education is the so-called digital storytelling. This term was coined as early as the 1980s and originally referred to the production of short films on topics from the lives of nonprofessional filmmakers (Robin 2008). In a broader understanding, however, it refers to the creation of a (short) story that is produced and presented using digital media and tools (Wu & Chen 2020). While the Dunedin example still used traditional oral storytelling, the use of storytelling in higher education now mostly takes place digitally. In addition to lecturers conveying content in story form, there is a growing use in the sustainability context where students use storytelling to explore topics and communicate their findings so that a nonspecialist audience can also follow the presentation, often working with elements of digital storytelling. For example, students at the universities in Tunis (Tunisia) and Hagen (Germany) produced short films in a 2-year project documenting people's lived experiences of climate change (Otto 2017). In just one semester, first-year students at Leuphana University created short films following a three-step process. At the beginning of the story, they described the present state from which a challenge emerges, and then understanding these challenges required a change in thinking and led to a journey of discovery, which ended the resolution of its conflict (Altekruse & Fischer 2020).

The use of digital storytelling can serve five purposes, as a recent review study by Wu and Chen (2020) shows. The most frequent type serves to help students better acquire scientific concepts and content (*appropriation*). The opening story about Johnnie's toothache fits this purpose. Digital storytelling can be used to give students autonomy over their learning process by allowing them to work with higher degrees of freedom (*agency*). A third purpose is to have students (e.g., in the context of final exams) consider and analyze the meaning of their learning experiences (*reflection*). Finally, Wu and Chen (2020) mention two critical variants of digital storytelling that contribute to questioning concepts (*reconstruction*) and to questioning one's own identity (*reflexivity*).

Storytelling can thus be used in very different ways in higher education, making it at times a difficult and confusing practice for lecturers to master (Ketelle 2017). In particular, the critical variants of storytelling, as mentioned by Wu and Chen (2020), in which one's own experience of injustice, discrimination, or oppression is made public to redress these grievances, seem to have been less widely used in sustainability-related teaching to date. Given the obvious social sustainability problems that can be addressed with this form of critical storytelling, it would be worth making this little-tested practice more fruitful in higher education teaching in the future (Astiz 2020).

5.2 SusTelling in Concepts of Teaching

5.2.1 Approaches

In some disciplines, stories are a central element in teaching, as are sustainability topics. But stories have not yet been widely used in higher education to address sustainability. One obstacle could be common disciplinary self-understandings. In economics, for example, there is a widespread view that learning content is "value-neutral", so that "normative" sustainability topics cannot be considered (see Moltan-Hill 2014). Similarly, narrative elements are often not considered appropriate or adequate teaching methods. Some lecturers may simply have basic problems with incorporating stories and/or linking sustainability issues to their subject content. Or, they may not even be aware that some seminar content can be taught in narrative form or what connections there are between disciplinary content and sustainability topics. In the following section, we want to show some ways to connect SusTelling to different disciplines.

Whenever practical work is done at the university, stories can help determine the perspectives and objectives for practical orientation. Short stories can serve a guiding function by raising questions such as: Why do we want to develop this technology? What problems in the world do we want it to solve? And, to give a more concrete example: Is it important for nuclear scientists to think about their responsibility if their research could be used to build a nuclear bomb?

Dürrenmatt reflected on this problem in his drama *The Physicists*. His book, or similar stories, could be integrated into a physics course to address the ethical implications of research. Furthermore, well-told short biographies of scientists can help connect data and facts with a life story. These personal histories offer students the opportunity to develop an understanding of what inspired researchers, what relevance their research had at the time, and what events helped their findings achieve a breakthrough. Feature films can also serve this function. One example is the film *The Insider*, starring Russel Crowe, based on real-life experiences.

A tobacco industry executive publishes research findings to alert the public to the health risks of smoking and dangerous production practices in the industry. Learners could watch a clip from the film and then discuss how they would act if they were in his situation.

In addition to the uses just described, short biographical elements, trailers of feature films, short stories, poems, and other narrative formats can also enrich teaching and learning in higher education settings in other ways. They can be used to summarize key points of a lecture or seminar and as mnemonic devices. Stories can also take up a larger space in the syllabus when several chapters or entire books are read during the course.

5.2.2 Case Studies

Two examples are presented later. The first example comes from the practical experience of one of the authors (Petra Molthan-Hill). It was taught several times in this form in the field of accounting in an elective. The learning objective was a deeper understanding of the conditions under which the so-called creative accounting becomes an illegal practice. The second example in the following section shows how science fiction films can be used in different disciplines.

A Novel for Learning Proper Accounting

The seminar design is focused on the book *Anatomy of Greed: The Unshredded Truth from an Enron Insider* by Bryan Cruver (2002). In this book, a former employee of the corporation recounts his experience of the company's collapse following its illegal accounting practices. The book is written in the style of a John Grisham thriller and is just as suspenseful. Illegal practices the author uncovers can be taken up and discussed in seminar sessions where proper accounting is compared to accounting that crosses the line. Students learn not only how accounting should be practiced but also how to identify illegal practices.

In this course, the book was also integrated into the final exam. Students were allowed to bring the novel into the final exam (open-book exam) and then had to answer various questions and relate the theoretical content they had learned to the scenes portrayed in the book. For example, they had to reproduce an ethical decision-making model and then use an example from the book to show how it would have led to different behaviors and outcomes.

A Science Fiction Trailer

Sustainability is an idea strongly oriented toward the future. Seminar sessions in various disciplines can use trailers or entire films from the science fiction genre to show possible future scenarios that relate to current social trends and problems in the area of sustainability. In their chapter in *Storytelling for Sustainability in Higher Education*, Jeffrey Barber et al. (2020) discussed several well-known feature films

TABLE 5.1 How the interplay of collective and individual as well as internal and external factors contribute to a climate catastrophe in the movie "The Day After Tomorrow" (adapted from Barber et al. 2020)

	Internal	*External*
Individual	**Psychological/Consciousness** Initial disbelief and skepticism over the warnings; then shock, confusion, and panic as reality of the catastrophe hits.	**Behavioral/Practices** Business as usual, followed by paralysis, then panic-stricken attempts to escape from floods, flying cars. People drowning, some people trying to help, others fleeing. Later efforts to simply survive, especially after the deep freeze. Efforts by father to search for and rescue his son.
Collective	**Culture/Worldviews** Despite scientific evidence, the dominant worldview is skeptical. A culture of high-carbon consumerism, resistant to calls for sustainable consumption and production. Neoliberal views about unencumbered economic growth prevail – until reality crashes preconceptions.	

Table showing how the interplay of collective/individual and internal/external factors in movie "The Day after Tomorrow" result in four different perspectives (psychological/consciousness, culture/worldviews, behavioral/practices, and social/systemic; adapted from Barber et al. 2020)

for classroom use and show how the film *The Day after Tomorrow* could be used to understand climate change from different scientific perspectives (Table 5.1). The film centers on a man who experiences a powerful storm in New York that ushers in a new ice age. The scenes in the film provide an opportunity for reflection on how the interplay of individual and collective, internal and external factors of human action contributed to the emergence of such a catastrophe. Psychological theories could be used to explain phenomena of repression and denial, or socio-critical approaches of sociology could be combined with philosophical reflections on neoliberalism.

5.3 SusTelling as a Method for Gaining Knowledge

5.3.1 Approaches

The focus of the current discussion on the use of storytelling in higher education revolves primarily around using narrative forms of communication to change understanding among students (for teaching concepts in higher education, see section 5.2) or other recipients and target audiences (for science communication, see section 5.4). In this respect, we can speak of using storytelling as an intervention. Next to this *interventionist* understanding, storytelling can also be used in the context of higher education to first understand a person's worldview instead of trying to change it. Such an *interpretive* use of storytelling has a long tradition in various research fields. In this section, we show that approaches from these research fields also hold valuable insights for educators in higher education.

A first step is to distinguish two ways of using storytelling to better understand how people think and talk about sustainability: as a method for data collection and as a method for data analysis. There is no hard and fast line dividing the two ways. Instead, they can overlap and merge into each other in research practice. Nevertheless, to systematize the uses of SusTelling in higher education, it seems helpful to distinguish the two ways as methods for gaining knowledge.

First, as a method for obtaining data, storytelling is used to verbalize internal processes. An example of this is the narrative interview approach in qualitative social research, in which interviewees are given ample space to unfold their narratives (McCormack 2004). In other varieties of qualitative interviewing, interviewees are encouraged by various stimuli (such as chronology, personalization, and dramaturgy) to describe their experience in episodes that exhibit characteristics of stories (Elliott 2005). Narrative forms, however, not only are used in interview situations but can also serve as a method of introspection and self-exploration in the tradition of autoethnography. Regarding sustainability issues, nature writing should be mentioned here, as it transforms scientific observations of nature into narrative writing, thus combining objective description and subjective experience of everyday encounters with nature (Moran 2014).

Second, as a way of analysis, SusTelling can draw on very different materials. On the topic of climate change, we find, among others, narrative cycles in climate reporting in the media (Gkiouzepas & Botetzagias 2017), representations of climate change in fictive literature (Trexler & Johns-Putra 2011), and position papers by climate policy groups (Blaxekjær & Nielsen 2015). This material can be examined to see what stories it tells about how authors relate to ideas and challenges of transitioning toward safe and just futures. Various narrative analytic methods (as opposed to, say, descriptive or persuasive) can help to capture and elaborate specific aspects of storytelling (Mertova & Webster 2019).

Common to both ways is that they use storytelling as a way of understanding and adding to students' method(olog)ical toolbox.

5.3.2 Case Studies

In the following sections, we present two case studies that illustrate how the use of storytelling in a higher education context can help to yield new insights and gain knowledge. The first example is from the North American context and addresses issues of social sustainability. The second example is a course that one of the authors (Daniel Fischer) piloted with Anna Sundermann (co-editor of this book) at a German university.

Testimonios: Exploring and Addressing Social Injustice Through Narrated Experience

Discrimination and oppression are incompatible with the values endorsed by the idea of sustainability, making social justice an important goal in sustainable development processes. The genre of *testimonios* emerged in Latin America in the 1970s as a way to expose, understand, and combat racism, xenophobia, and other forms of discrimination and oppression (Reyes & Curry Rodríguez 2012). Testimonios are narrated experiences in which a narrator presents their own experience of social injustice, after reflecting on them to make them public. That is, it highlights grievances and demonstrates solidarity with others affected by similar injustices to redress these wrongs. This intentionality distinguishes testimonios from other forms of self-documented experience (e.g., diaries). Testimonios always have a pedagogical *and* an activist element grounded in emancipation, empowerment, and collective intervention, as well as, where appropriate, healing processes (Delgado Bernal et al. 2017). More recently, testimonios have been understood as an approach to generating knowledge. In testimonios, experiential and reflective knowledge about structures and processes of social inequality emerges, which claims legitimacy in its own right (DeRocher 2018). In other words, the perspectivity of authentic experience is not seen as something to be overcome in the course of scholarly translation and analysis but as a form of valid knowledge produced by people with lived experiences of oppression. Some researchers see testimonios as a distinct form of knowledge production that deliberately sets itself apart from Western research traditions (Huber 2009).

Testimonios are used in higher education to explore and collectively raise awareness of experiences of discrimination and oppression that students have experienced. Work with testimonios has become widespread, particularly in social justice education in North America, where often experiences of students of color are addressed in the form of testimonios (Sosa-Provencio et al. 2019). However, their use in the field of Higher Education for Sustainable Development is still in its infancy.

Greening Duckburg: Sustainability in Comic Books

Media articles, interviews, and policy documents are among the materials that are often subjected to narrative analysis. But what can stories in comic strips tell us about how sustainability issues enter into and are framed in, contemporary discourses? First-year students explored this question as part of the seminar *Greening Duckburg – Comics as a Medium of Sustainability Communication*, offered in the winter semester of 2015–2016 at Leuphana University. In a longitudinal study, they examined the stories to determine the extent to which sustainability is addressed in this comic strip and how the topic of sustainability in the Duckburg stories has changed over time. On the basis of their analysis of a large number of stories, students also developed character studies for individual figures in Duckburg and discussed typical ways of thinking and behaving in dealing with sustainability issues.

The seminar is an example of how storytelling can be used to analyze verbal and visual communication on sustainability issues. Even though elaborate methods of narrative analysis were not used in this seminar for first-semester students, the approach did allow them to gain insights into the presence and framing of sustainability issues even in narratives that superficially did not seem to have any immediate relevance to sustainability.

5.4 SusTelling as a Method of Science Communication

5.4.1 Approaches

Learners in every discipline in higher education will need to communicate their findings in their future careers. This may consist in enumerating factual information, but often facts must first be interpreted to proceed to the next stage of work. This involves condensing factual information – sometimes unconsciously – in narrative form to collectively develop stories. One example is climate communication, where researchers who had long presented only factual information began to interpret them for a broader media audience by talking about climate change, global warming, or, most recently, the climate crisis. Not infrequently, these terms were used together with brief dystopian descriptions of their effects – including doomsday scenarios. It is now recognized that this form of scientific presentation has only a limited effect in engaging the broader population and decision-makers in a discussion of the facts and persuading them to initiate decisive steps to avert the catastrophic effects of unchecked climate change. Various research projects show that this engagement can only be achieved if the factual information is presented with either positive identification figures (the so-called Greta effect) or positive narrative elements (Baden 2020).

For university teaching, we might therefore ask to what extent, for example, spaces for reflection on common narratives in the social application of science (as illustrated here with climate change, but also relevant for many other subjects) should be integrated into science curricula. Learners could also be motivated to present scientific findings in ways that are appealing and understandable to the media and other decision-makers. This requires the competency not only to present scientific facts in everyday language but also to distinguish clearly between facts and evaluative judgments (Habermas 1984). The presentation of contextualized facts in narrative formats and a conscious decision to do so in the form of SusTelling (see Chapter 1) could usefully complement the canon of classical teaching and learning objectives in science. This approach is also relevant for the social sciences and humanities if they want to make their research results accessible to a wider audience. Interdisciplinary projects, in particular, lend themselves to this approach. In a joint course at Nottingham Trent University, for example, biology students are learning how to write about the natural sciences in a journalistic way together with students studying journalism learning to distinguish fake news from real news in the natural sciences and to interpret natural science facts appropriately (https://bit.ly/2VhNcIA).

5.4.2 Case Studies

The following two case studies (story blogging and poems as narrative forms of examination) demonstrate how storytelling can be used in higher education to foster science communication skills in students.

Story Blogging as a Form of Examination

Blogging as a form of Internet-based writing or filming is now an established form of science communication. Puntha et al. (2020) provide several examples to illustrate how story blogging can be fruitful as a teaching method and then as a form of examination in higher education. Students of history, for example, could report on historical excavations and awaken the interest of a wider audience. In the social sciences, on the other hand, students could examine a controversial policy measure to restrict migration and immigration by writing a blog from the perspective of a humanitarian aid organization and then from the perspective of a member of a right-wing populist party. Real blog posts could be analyzed to see to what extent their authors use reliable sources and how they arrive at their opinions. Students could develop criteria for recognizing fake news and how to distinguish between fact and fiction. Authenticity is another concept that lends itself to a seminar dealing with the topic of blogs.

The analyses of blogs, YouTube videos, TED Talks, Instagram posts, and other items on social media offer a variety of opportunities to think about what conditions are needed for communication to be effective and then develop subject-specific communication strategies. The goal here would be to nurture a more conscious use of social media and an awareness of their narrative elements, including insights into how these narrative elements are used to influence attitudes and/or move readers to action, especially in the context of sustainability.

Regarding assessment criteria, it should be noted that it is not the substance of personal attitudes that should be assessed but the student's ability to reflect on them, place them in an appropriate context, and provide detailed and comprehensible evidence. Puntha et al. (2020) provide assistance on how to formulate assessment criteria appropriate for blogging and other narrative forms of assessment.

Poems as Final Papers

Stories such as Heinrich Böll's anecdote about a fisherman that a tourist is trying to persuade to fish more before resting (see Chapter 1) can be used in courses in a wide variety of disciplines, for example, to stimulate reflections on meritocracy or the necessity and quality of economic growth. In the same way, students can be motivated to "condense" their findings in lyrical form.

Students of all disciplines at Nottingham Trent University in England have had the opportunity to earn a Sustainability in Practice Certificate by writing a story (https://bit.ly/2VhNcIA), added in 2018 as an assessment option to the online course run since 2012 (Molthan-Hill et al. 2016). The online course asks students to choose between three topics: How can we feed the world sustainably? How can we produce sustainable energy for the world? and How can we produce sustainable clothing for the world? After working through four teaching units (individual, disciplinary, interdisciplinary, and then international solutions), they decide whether to submit their discipline-based sustainable solution as a video, presentation, artifact, project design, or story/poem. Contrary to the expectations of the program organizers, who had expected that students taking literature studies would submit stories or poems, several poems came from students in the business school. The lecturer in economics who received one of the poems for assessment reported that this was the first time in her life she had been deeply moved by an exam paper. The student had managed to summarize sustainable issues in this poem in a way that made the significance of the economic ills he had analyzed more "tangible" while evoking empathy. At the same time in an accompanying chapter, he supported each lyrical device he had used (e.g., metaphors) to convey the content with references to statistics and theories.

5.5 Conclusion

In this chapter, we have proposed that sustainability-related storytelling in higher education can be understood from three different perspectives: as a possibility of alternative teaching–learning formats, as a method for gaining knowledge, and as a practice of science communication. As the different case studies show, storytelling can make valuable contributions to expanding and complementing the methodological repertoire of teaching in higher education. However, the three perspectives discussed in this chapter also show that the use of storytelling should be well considered if pitfalls are to be avoided and storytelling is to unfold its potential. We want to conclude by highlighting three insights that seem particularly significant to us: the importance of the functions of storytelling, dealing with contingency in storytelling, and the largely undeveloped potential of storytelling approaches in higher education.

The first insight emphasizes the different purposes – instrumental and critical emancipatory – that the use of storytelling in higher education can serve. Narrative forms can be used instrumentally to communicate academic knowledge and ideas so that they can be better understood, remembered, and applied. This includes forms of storytelling production in which learners, through active storytelling, are required to transform their knowledge and organize it in ways different from what is usually required of them in traditional exam situations. This instrumental use of storytelling thus aims at improving teaching–learning interaction in higher education and should be distinguished from storytelling in which social reality is expressed, interpreted, and changed. The approaches of critical and emancipatory storytelling rooted in liberation pedagogy (Freire 2000) should be mentioned here, as well as the more activist forms of science communication aimed at initiating and supporting social change processes toward sustainable development (Davis et al.

2018). In this respect, teachers should be clear about their goal for using storytelling in their teaching. Furthermore, they need to provide opportunities for students to reflect on storytelling, what distinguishes it from other forms of communication, and how it works or could work.

A second insight points to an essential difference between storytelling and classical forms of presenting scientific findings, namely, that the use of stories is always connected with contingency. Stories allow for different readings, associations, and interpretations – which can be potentially enriching. Contingency can encourage students to develop more complex forms of knowledge production, representation, and communication. At the same time, the contingency of stories lies always in tension with specific teaching and learning goals in higher education, that is, a story can always be understood differently than the lecturer intended. In dealing with this tension, it is important not to use stories only to achieve very specific teaching goals but to preserve their contingency and use it productively.

A third insight lies in encouraging experimentation with forms of storytelling that have been little used in higher education. We see critical storytelling as particularly promising in this regard. Stories can help us not only to describe and understand unsustainable conditions but also to create community and solidarity to change such conditions. They also allow academic knowledge to be related to lived experience, which is crucial for developing competencies. Critical forms of storytelling could be fruitful for all three perspectives outlined in this chapter. This is a task for further research and experimentation in higher education pedagogy.

Finally, we would like to venture a hopeful outlook. Despite the pioneering work that has already been done, storytelling in higher education is still a fairly young and new practice. Networks of researchers and practitioners sharing their experiences with SusTelling in higher education are particularly important to the development of this practice. There is a range of organizations and resources that allow interested educators to further develop the promising and growing practice of storytelling in higher education (for a review, see Wall et al. 2020).

References

Altekruse, J., & Fischer, D. (2020). Digitally-enhanced learning through collaborative filmmaking and storytelling for sustainable solutions. In P. Molthan-Hill, D. Baden, T. Wall, H. Puntha, & H. Luna (eds.), *Storytelling for Sustainability in Higher Education: An Educator's Handbook* (pp. 103–116). London: Routledge.

Astiz, M. F. (2020). Storytelling in the higher education classroom: Why it matters. *College Teaching, 68*(4), pp. 187–188.

Baden, D. (2020). Which work best: Cautionary tales or positive role models? In P. Molthan-Hill, D. Baden, T. Wall, H. Puntha, & H. Luna (eds.), *Storytelling for Sustainability in Higher Education: An Educator's Handbook* (pp. 88–100). London: Routledge.

Barber, J., Onthank, K., Wall, T., Johnson, N., & Mackenzie, A. (2020). Examining sustainability challenges using science-fiction film scenarios. In P. Molthan-Hill, D. Baden, T. Wall, H. Puntha, & H. Luna (eds.), *Storytelling for Sustainability in Higher Education: An Educator's Handbook* (pp. 117–129). London: Routledge.

Blaxekjær, L. Ø., & Nielsen, T. D. (2015). Mapping the narrative positions of new political groups under the UNFCCC. *Climate Policy, 15*(6), pp. 751–766.

Cruver, B. (2002). *Anatomy of Greed: The Unshredded Truth From an Enron Insider*. New York: Carroll & Graf Publishers.

Davis, L., Fähnrich, B., Nepote, A. C., Riedlinger, M., & Trench, B. (2018). Environmental communication and science communication – conversations, connections and collaborations. *Environmental Communication: A Journal of Nature and Culture, 12*(4), pp. 431–437.

Delgado Bernal, D., Burciaga, R., & Flores Carmona, J. (2017). *Chicana/Latina Testimonios as Pedagogical, Methodological, and Activist Approaches to Social Justice*. New York: Routledge.

DeRocher, P. (2018). *Transnational Testimonios: The Politics of Collective Knowledge Production (Decolonizing Feminisms)*. Seattle: University of Washington Press.

Elliott, J. (2005). Listening to people's stories: The use of narrative in qualitative interviews. In J. Elliott (ed.), *Using Narrative in Social Research* (pp. 17–35). London: Sage.

Freire, P. (2000). *Pedagogy of the Oppressed* (30th anniversary ed.). New York: Continuum.

Gkiouzepas, G., & Botetzagias, I. (2017). The narrative cycle of climate change in the Greek newspapers, 2001–2008. *Environmental Communication, 28*(3), pp. 1–18.

Glonek, K. L., & King, P. E. (2014). Listening to narratives: An experimental examination of storytelling in the classroom. *International Journal of Listening, 28*(1), pp. 32–46.

Habermas, J. (1984). *The Theory of Communicative Action: Reason and the Rationalization of Society*. Cambridge: Polity Press.

Huber, L. P. (2009). Disrupting apartheid of knowledge: Testimonio as methodology in Latina/o critical race research in education. *International Journal of Qualitative Studies in Education, 22*(6), pp. 639–654.

Ketelle, D. (2017). Introduction to the special issue: What is storytelling in the higher education classroom? *Storytelling, Self, Society, 13*(2), pp. 143–150.

Kieser, J., Livingstone, V., & Meldrum, A. (2008). Professional storytelling in clinical dental anatomy teaching. *Anatomical Sciences Education, 1*(2), pp. 84–89.

Manteaw, B. O. (2020). Consumer privileges and the sociocultural dimensions of production and consumption. In P. Molthan-Hill, D. Baden, T. Wall, H. Puntha, & H. Luna (eds.), *Storytelling for Sustainability in Higher Education: An Educator's Handbook* (pp. 193–204). London: Routledge.

McCormack, C. (2004). Storying stories: A narrative approach to in-depth interview conversations. *International Journal of Social Research Methodology*, 7(3), pp. 219–236.

Mertova, P., & Webster, L. (2019). *Using Narrative Inquiry as a Research Method. An Introduction to Critical Event Narrative Analysis in Research, Teaching and Professional Practice*. London: Routledge.

Moltan-Hill, P. (2014). The moral muteness of managers: An Anglo-American phenomenon? German and British managers and their moral reasoning about environmental sustainability in business. *International Journal of Cross-Cultural Management, 14*(3), pp. 289–305.

Molthan-Hill, P., Baden, D., Wall, T., Puntha, H., & Luna, H. (eds.) (2020). *Storytelling for Sustainability in Higher Education: An Educator's Handbook*. London: Routledge.

Molthan-Hill, P., Dharmasasmita, A., & Winfield, F. M. (2016). Academic freedom, bureaucracy and procedures: The challenge of curriculum development for sustainability. In J. P. Davim & W. Leal Filho (eds.), *Challenges in Higher Education for Sustainability* (pp. 199–215). Cham: Springer.

Moran, J. (2014). A cultural history of the new nature writing. *Literature & History, 23*(1), pp. 49–63.

Otto, D. (2017). Lived experience of climate change – a digital storytelling approach. *International Journal of Global Warming, 12*(3/4), pp. 331–346.

Puntha, H., Touboulic, A., & Wall, T. (2020). Use of storytelling within assessment to support sustainability literacy. In P. Molthan-Hill, D. Baden, T. Wall, H. Puntha, & H. Luna (eds.), *Storytelling for Sustainability in Higher Education: An Educator's Handbook* (pp. 168–190). London: Routledge.

Reyes, K. B., & Curry Rodríguez, J. E. (2012). Testimonio: Origins, terms, and resources. *Equity & Excellence in Education, 45*(3), pp. 525–538.

Robin, B. R. (2008). Digital storytelling: A powerful technology tool for the 21st century classroom. *Theory into Practice, 47*(3), pp. 220–228.

Singer-Brodowski, M., Grossmann, K., Bartke, S., Huning, S., Weinsziehr, T., & Hagemann, N. (2018). Competency-oriented education for sustainable development. *International Journal of Sustainability in Higher Education, 19*(7), pp. 1299–1316.

Sosa-Provencio, M. A., Sheahan, A., Füntes, R., Muñiz, S., & Prada Vivas, R. E. (2019). Reclaiming ourselves through Testimonio pedagogy: Reflections on a curriculum design lab in teacher education. *Race Ethnicity and Education, 22*(2), pp. 211–230.

Stoltenberg, U., & Burandt, S. (2014). Bildung für eine nachhaltige Entwicklung [Education for sustainable development]. In H. Heinrichs & G. Michelsen (eds.), *Nachhaltigkeitswissenschaften [Sustainability Sciences]* (pp. 567–594). Dordrecht: Springer.

Trexler, A., & Johns-Putra, A. (2011). Climate change in literature and literary criticism. *Wiley Interdisciplinary Reviews: Climate Change, 2*(2), pp. 185–200.

Wall, T., Puntha, H., Molthan-Hill, P., Kemp, N., Puntha, S., & Baden, D. (2020). Conclusion – Stepping into sustainable futures: Actions, developments and networks. In P. Molthan-Hill, D. Baden, T. Wall, H. Puntha, & H. Luna (eds.), *Storytelling for Sustainability in Higher Education: An Educator's Handbook* (pp. 386–398). London: Routledge.

Wu, J., & Chen, D.-T. V. (2020). A systematic review of educational digital storytelling. *Computers & Education, 147*, 103786.

6

STORYTELLING FOR SUSTAINABILITY AND CONSTRUCTIVE JOURNALISM

A Perfect Pair?

Maren Urner

In this chapter, I argue that Storytelling for Sustainability (SusTelling) and constructive journalism can be combined. The combination of both approaches automatically addresses fundamental challenges that journalism is currently facing. Specifically, I discuss constructive journalism as an opportunity for journalism, especially with regard to narrative approaches to sustainability, without claiming to give a comprehensive analysis of the journalistic perspective in combination with SusTelling.

In 2016 I, Maren Urner, co-founded *Perspective Daily* (perspective-daily.de), the first ad-free online magazine for constructive journalism in the German-speaking world. The decision to co-found a media company after completing my doctorate in neuroscience was based on various shortcomings I observed in journalism. In terms of content, I was primarily concerned with the failure of media in the area of sustainability. Methodologically, I was concerned about the excessive focus on the negative and the neglect of a future-oriented perspective, as well as the lack of factual competency among media professionals (Guo & Volz 2021; Willnat et al. 2013).

As a cognitive and neuroscientist, my media analysis focuses primarily on the effect of news coverage on perceptual, attentional, and memory processes in recipients. This places my work at the intersection of journalism, neuroscience, and media psychology – since 2019 as a professor of media psychology. This chapter is about these interfaces, with the focus on constructive journalism.

In 2016, Austrian journalist Claus Reitan noted, "There have never been so many starting points for a journalism dedicated to sustainable development" (Reitan 2016, p. 148; translated by author). Under the influence of media change and digitalization and in the discourse on the newly proclaimed Age of Humans, the Anthropocene, there are many calls for a change in journalistic reporting to stimulate discussion, support people in developing an opinion, and facilitate individual

DOI: 10.4324/9781003326144-10

decision-making. Such a sustainability-oriented journalism faces numerous challenges in this regard (Berglez et al. 2017). One of the biggest is to get through to media users with sustainability topics that are often complex. After all, attention has become the most important and arguably the most valuable resource in the age of smartphones, mobile Internet, and constant accessibility (The Economist 2017). Websites, networks, and ads fight for clicks and scrolls. Even journalistic providers can no longer avoid this battle (Wu 2016). So they try to entice, seduce, and immerse recipients. Storytelling plays an important role in this. Whether it's a campfire or a family party, a news or a lifestyle magazine, a conversation with good friends, or the latest series on a streaming service – writers almost always choose the form of a story.

In the battle for attention, finding the right amount of story is no easy task. It can sometimes happen that the line between genuine emotional closeness and fiction becomes blurred. While the focus of storytelling was initially still on narrative elements describing external appearances, moods, and situations, the desire to get as close as possible to protagonists has sometimes taken on absurd dimensions. This was most recently demonstrated by the Relotius affair,[1] which in view of invented interviews and stories not only became a proper scandal for the news magazine *Der Spiegel* but also triggered a debate on credibility in the perhaps excessive use of stories in journalism in general and the journalistic form of reporting in particular (Lünenborg & Medeiros 2021).

The human brain likes stories because they are emotional. Simply put, the more emotional content is, the better it is stored in the brain (LaBar & Cabeza 2006). This is why people who are old enough still remember very clearly what they were doing on 9/11.

This chapter is about the combination of SusTelling – that is, journalistic storytelling focused on sustainability – with so-called constructive journalism. In the spirit of storytelling, some theoretical background to this combination is based on a concrete example.

6.1 When the Story Becomes Real: A Practical Example

Suddenly sirens are wailing. Police officers and armed soldiers block major intersections and roads in the cities, and within a few hours the streets are deserted. People rush on foot to the supermarkets to hoard food. Laden with water bottles, cans of food, and a good dose of uncertainty, they hurry home. Presenters and commentators babble wildly on all media channels. The cell phone networks are overloaded. Parents text their children: "Are you okay? Are you safe? Do you know what's going on?" Breaking news in large red letters flickers across the screens: "Government declares state of emergency".

So begins the magazine article "In 5 Years a State of Emergency Will Be the Only Way Forward for Germany", which I published with Felix Austen at *Perspective Daily* on December 28, 2016 (Austen & Urner 2016) – shortly after the Australian city of Darebin declared a climate state of emergency on December 5,

2016 (CACE n.d.). It was not until May 2, 2019, that Constance became the first German city to declare a climate state of emergency (Knodt 2019). This state of emergency was then declared by nearly 70 German cities, the EU Parliament (2019), and numerous actors around the world by the end of the same year.

In the same year, Oxford Dictionaries chose "climate emergency" as its word of the year. Their reasoning? Use of the word had increased 100-fold in the previous 12 months, lending a "greater sense of urgency" to the way we talk about the climate (Zhou 2019). Climate emergency is defined in this context as "a situation in which immediate action is needed to reduce or end climate change and prevent serious and permanent damage to the environment" (Oxford Dictionaries 2022).

It is this immediate action that is the subject of the magazine article at *Perspective Daily*, which after the opening scene continues its description of the fictive state of emergency. In a speech printed in the magazine article, Chancellor Angela Merkel addresses the German population and announces that a 12-month emergency program is now coming into effect in order to secure Germany's free and democratic constitutional order in the long term. The narrative continues with the three phases of the emergency program, from the 2-month Phase 1 ending the status quo, to the 4-month Phase 2 of rebuilding and changing the five major areas of life finance, transportation, food, housing, and work, to the 6-month Phase 3 of habituation, a test run for the new order. The text concludes with a description of "life after the state of emergency", which also takes place in the future (on May 1, 2018).

Of course, this chapter is not meant to suggest a possible causal relationship between the use of climate emergency as a story for a 2016 piece of journalism and the subsequent success of the term and the concept behind it. Instead, using the example of climate emergency and an early, at the time still futuristic, journalistic article using storytelling, I describe in this chapter how constructive journalism can be applied in the field of SusTelling – with practical aspects embedded or explained in the theoretical framework.

6.2 SusTelling in Constructive Journalism

The impetus and inspiration for this unusual magazine article about a fictitious climate emergency in Germany came from an earlier piece by the American climate activist Bill McKibben (2016). In his August 2016 op-ed "A World at War", the author and founder of the international climate protection organization 350.org compares the extent of the threat posed by man-made climate change and the transgression of planetary boundaries[2] with the rise of fascism and Adolf Hitler's actions during the Nazi era. He considers the current threat to humanity from man-made climate change a greater threat and calls for a global mobilization of humanity to direct all action toward winning the "battle" against the threats posed by climate change (in the broadest sense) so we can still avert the worst consequences.

Bill McKibben cites the American economy at the time of World War II as an example of a goal that united all Americans: defeating fascism. Factories were repurposed to produce war machinery. Instead of refrigerators and pick-up trucks, tanks, bullets, and fighter jets were produced. Today, this would mean converting the production of diesel engines into solar panels and turning highways into bicycle lanes. Interestingly, due to the Corona pandemic, a similar conversion can be observed in many sectors of the economy. Instead of T-shirts and pants, masks are being produced, and instead of car engines, ventilators are being built. The goal of the magazine article "In 5 Years a State of Emergency Will Be the Only Way Forward for Germany" was to tell a story about a near future of a Germany that prioritizes preventing the consequences of climate change. The challenge was to find a balance between storytelling and realism by referring, for example, to the legal status of emergency laws and the design of a completely sustainable economy. At the same time, in the spirit of constructive journalism, the goal was to create a vision that would give recipients a hopeful look ahead. In order to better understand this approach, an introduction to constructive journalism using the article as a case study follows.

6.3 Constructive Journalism: A Case Study

6.3.1 Brief Background

The purpose of constructive journalism is not to reinvent journalism but rather to do better justice to its core tasks than is usually the case at present. In very general terms, one of these core tasks is to provide the public with socially relevant information in order to contribute to the formation of public opinion (Kovach & Rosenstiel 2007). Journalism is therefore often described as the fourth estate in a democracy, with a certain watchdog function (McNair 2009).

Probably one of the most important reasons for journalism's failure to achieve this goal is the strong negativity bias on the part of both journalists and recipients, which gives most people an overly negative view of the world.[3] For media professionals, this preference ensures that crises, wars, and scandals are especially likely to make it into the headlines, and among recipients, a presumably evolutionary preference for the negative leads to increased consumption of precisely this kind of information. However, the media have been criticized "for publishing too much bad news for more than half a century" (McIntyre 2016, p. 223).

This tension gave rise to the basic idea of constructive journalism (Lough & McIntyre 2021). Summarized in one sentence, constructive journalism goes beyond the classic '5Ws+H'-questions of journalistic work of What? Where? When? Who? How? Why? by always asking: What now? (or, a little nicer if somewhat longer: What happens next?). It is crucial that this is not formulated as a last-on-the-list question but as one that permeates all of the work that journalists do. With the what-now-question in mind, journalists are able to take a more expansive approach to their work. This begins with choosing the protagonists, images, and sources and ends with the structure of the magazine article and the choice of words and phrases.

Constructive journalism is often thought to have originated in Scandinavia, in Denmark more precisely, but this view is short-sighted. Since in recent years parallel developments and currents have emerged internationally, especially in the United States and the Netherlands, all of which can be considered a constructive approach to journalism. For the first academic study worldwide situating the concept in the field, interested readers are referred to the paper by Karen McIntyre and Cathrine Gyldensted from 2017. Without going into further theoretical details here, at least one other important aspect seems worth mentioning with regard to the connection to SusTelling. This is especially important with regard to approaches from the North American and British regions, where the term "solutions journalism" is often used. The distinction between solution or solution-oriented journalism and constructive journalism is quickly explained with the help of simple set theory. Solution journalism is a subset of constructive journalism. It discusses – in response to the what-now question – actual or proposed solutions, where a constructive report does not have to present concrete solutions but can "simply" consist of providing an interpretation of the issue.

Fundamental to the constructive approach – no matter what type – is the changed way of thinking that accompanies the what-now question and casts light on a different, often new view of the world and one's own role and possible effect in it. This way of thinking has been aptly summarized by American psychotherapist Steve de Shazer, who summed up the shift in thinking as follows: "Problem talk creates problems, solution talk creates solutions".

In the following, the three core aspects or ingredients of constructive journalism – theory, form, and craft – are briefly introduced, and the theoretical principles of sustainability-related storytelling are applied to the magazine article "In 5 Years a State of Emergency Will Be the Only Way Forward for Germany".

6.3.2 Theory: A New Way of Looking at the World

At the core of the theory behind constructive journalism are insights from psychology, specifically from behavioral and positive psychology. Simply put, it is about the realization that people can enter a state of learned helplessness in the face of repeated confrontation with inescapable situations, for example, unsolvable tasks (Maier & Seligman 1976). Transferred to media consumption – problem-oriented reporting without the additional what-now-question – this feeling of helplessness can be fostered in recipients and accompanied by chronic stress and other negative health consequences. Constructive journalism also builds on numerous insights from positive psychology. The two co-founders of this fledgling psychological subdiscipline, Martin Seligman and Mihaly Csikszentmihalyi, describe the aim of positive psychology as an attempt "to begin to catalyze a change in the focus of psychology from preoccupation only with repairing the worst things in life to also building positive qualities" (Seligman & Csikszentmihalyi 2000, p. 5). In such perspective, positive psychology is "nothing more than the scientific study of ordinary human strengths and virtues", focused on the "average person", exploring "what works, what is right, and what is improving" (Sheldon & King 2001, p. 216). An example is the well-researched finding that positive emotions increase our willingness and ability to act, as well as creativity (Fredrickson 2013). Conversely, this means that journalism's core mission of fostering an informed and action-oriented public can be enhanced through constructive journalism.

This is precisely where the solution-focused mindset, already introduced by the quote from Steve de Shazer, comes in. By asking the overarching what-now-question, a different substantive discussion on the issues emerges. This solution-oriented discussion can in turn promote future-oriented thinking and action – the basic prerequisite for sustainability. At the same time, the resulting perspective provides a more complete picture of the world, since – as described – the negative focus of reporting leaves people with a distorted, overly negative view of the world on average. It is important to note that the task of media professionals is neither the development of solutions for a sustainable future nor greenwashing.

The changed viewpoint that follows asking the additional what-now-question is not easy, especially for journalists with years of professional experience, because old habits resist seeing things from a constructive perspective. In order to escape old patterns of thought and action, one quality is particularly important: curiosity. With a desire for new perspectives, a constructive view of the world comes almost by itself. This curiosity can help to internalize the three essential insights of this "new way of looking at the world". They are based on the application of research findings from positive psychology and cognitive science. They are briefly explained with reference to the magazine article "In 5 Years a State of Emergency Will Be the Only Way Forward for Germany".

Insight 1: Constructive Criticism Is More Than Pointing Out Wrong-Doings

This insight is central to understanding that solution-oriented thinking does not ignore problems but – quite the opposite – goes one step further than a problem-oriented approach. Put another way, *because* a situation is problematic there is a need for solution-oriented approaches. Constructive criticism is most certainly not about ignoring challenges and problems.

Applied to the state of emergency article, this means that humanity is at a historically unique crossroads where it must develop sustainable solutions if its own species is to continue its existence on planet Earth. In the Anthropocene era, humanity itself has become one of the most important influences on biological, geological, and atmospheric processes on Earth. So instead of remaining politically and socially stuck in the status quo, getting lost in cynicism, and complaining about what's wrong, the magazine article offers constructive criticism. It uses all the elements of storytelling (see Chapter 1) to describe a future that is far removed from the status quo – but one that is near in time and legally possible.

The most important element of constructive criticism in the magazine article is the so-called marix program (made up from my first name, Maren, and that of my colleague Felix Austen), which outlines a concrete roadmap to a sustainable future for the five major areas of life: finance, transport, food, housing, and work. This program includes the two normative requirements of SusTelling. The article aims to educate recipients – with regard to both the constitutional basis of a possible state of emergency in Germany and the technical aspects of what a sustainable future for the five major areas of life might look like in concrete terms. In order to support change toward sustainable development and to promote the ability of people to act in a self-determined and competent manner for the benefit of sustainable development (see Chapter 1), the magazine article emphasizes that the technology and knowledge needed for a sustainable way of life is already available. It is only the "wisdom" to use them that is still lacking. This is exactly where the recipients come in, who should be encouraged by the story to recognize their own potential impact and their own importance in the change toward a sustainable way of life and economy. This effect is directly related to insight 2.

Insight 2: Every Piece of Information We Perceive Has Influence

In many journalists' minds, there is still a misconception that simply pointing out wrongs will cause people to change them. As mentioned earlier, however, focusing on negatives can, in extreme cases, have the exact opposite effect, the so-called learned helplessness. This can ultimately result in a complete turning away from the challenges of our time and entering a state of passivity.

Another misconception is one frequently cited as a standard of journalistic quality, namely, to report objectively on events or rather to provide an objective overview of world events (Schudson 2001). What may initially sound like a laudable goal is simply an impossibility. First of all, this is due to the fact that journalists always make choices, whether about their subject matter or the people they select to interview and their other sources or in the design of their report. Yet it is exactly this aspect, as explained earlier, that is the basis for the definition and at the same time meaning of constructive journalism (Urner & Langeslag 2018). Moreover, every piece of information that the human brain processes creates a change that is measurable on a physiological level (e.g., neuronal) and on a psychological or behavioral level.

In general terms, and based on the findings of neuroscience, any information that is sent and received stimulates changes in the brain of the sender as well as in the brain of the receiver. This is why it is also misleading to speak of our brain or the human mind as a computer or a hard drive. The human brain always links new information with already existing memories and neural activity patterns, so that simple storage and deletion like on the hard disk of a computer is impossible. To put it constructively, instead of proclaiming the supposed objectivity of constructive journalism, its representatives call for more openness and honesty, for example, about research experiences and one's own values. Only in this way do recipients have a realistic chance of understanding the arguments and interpretations of a journalistic article.

Words such as "freedom" and "democracy" in the magazine article "In 5 Years a State of Emergency Will Be the Only Way Forward for Germany" should be mentioned here as examples. Even people sharing similar political convictions have different associations with these words. For some, the greatest possible freedom may consist in the absence of rules, and for others the exact opposite may be the case, because they are convinced that only through rules can the most important freedom – namely, the right to life – be secured in the long term. It is precisely in this area of tension between values and concepts of a good life that the fictitious state of emergency takes place, thus raising normative questions once again. First and foremost, the question: What constitutes a good life? This basic question leads to insight 3.

Insight 3: Putting an End to Black-and-White Thinking

In very few cases is the world black and white – especially not when it comes to complex interrelationships such as creating a sustainable society. Nevertheless, journalists often act as if there is always only one opposing view, usually in a personified form. The classic example is the ongoing discussions about man-made climate change, in which a so-called climate change denier is brought on stage to take the supposed counter-position to the climate scientist. The fact that there is an unprecedented scientific consensus on the subject is ignored.

Black-and-white thinking usually starts on an even more basic level, because – following on from insight 2 – people are assumed to understand the same things by a term. Analysis takes place too rarely, and correlations are passed off as causalities and blanket statements à la "coffee is healthy/unhealthy" serve as headlines. Without becoming too methodical at this point, constructive journalism is about better depicting and communicating complexity. This means recipients are expected to do more than just skim over bold headlines. Here, too, a more complete portrayal of the world plays a role in stimulating reflection and rethinking positions. Currently, the success and potential of such approaches to use journalism to stimulate discussion about complexity are being investigated in various journalistic and scientific projects (Alfter n.d.).

The shades of gray in the magazine article "In 5 Years a State of Emergency Will Be the Only Way Forward for Germany" consist in, for example, the fact that the matrix program for a sustainable social order does not present any "simple" solutions. In other words, it avoids black-and-white thinking to examine nuanced concepts in the five major areas of sustainable living. These concepts are merely outlined in this chapter, as it would otherwise be too long, but the objectives are clearly defined in each case. Their normative character is also very clear here, as a discussion about objectives cannot take place without a discussion about values.

6.3.3 Form: With All Our Senses and in Exchange With Others

There are a number of different ways the theory behind constructive journalism can be applied to journalistic reporting. With the subject matter of this book in mind, a special focus will be placed here on the application of storytelling, again using the magazine article "In 5 Years a State of Emergency Will Be the Only Way Forward for Germany".

As mentioned in the introduction, the human brain "likes" stories because they are more emotional than factual descriptions of contexts and events. They are not only processed by neural networks more intensively but also stored more securely. In other words, plot and personification enable recipients to engage with a topic much more intensively. Like experiences in "real life", stories are especially "real" when we experience them with different senses. An exchange with other people during or after a particular experience also ensures increased engagement and more intensive memory traces. Based on the individual aspects of SusTelling (see Chapter 1), the three aspects of story, interactivity, and exchange are examined later.

The plot of the magazine article has a clear structure of characters and plots given by a clear temporal framework. The year begins with the declaration of a state of emergency and ends with a vivid description of life after a successful transition to a sustainable social, political, and economic world. The article does not use the classic hero's journey, and only the real-life German Chancellor Angela Merkel appears as a protagonist when making her fictive speech. Through the pictorial description and additional elements (tonality and modality, see Chapter 1), the recipients themselves are meant to become protagonists witnessing the state of emergency. Personalization thus takes place by addressing the recipient in general terms, which are made as vivid as possible through a narrative style of writing (e.g., passages such as: "On social media, at the kitchen table and among the regulars at a bar, speculation is running wild, and the first conspiracy theories are circulating. The recommendations of the federal government last August to stockpile food and drinking water for at least two weeks are taking on a new urgency. What the hell is going on?").

After the opening scene and Merkel's speech, dramaturgy, chronology, and context are clearly oriented to the year-long transition to sustainable economic activity and coexistence. An attempt is made to strike a balance between narrative elements such as the mood on the nation's streets and informative, educational passages. The context is always clearly marked and at the same time should allow a relatively seamless transition between fiction and information.

An arc of tension (stylistics) is built up, especially at the beginning with an initially cryptic description of what is happening, based on a kind of end-time scenario intended to create certain images and interests in the minds of the recipients. The tonality, with colloquial expressions and an increased use of adjectives, is supported by illustrations created especially for the article. Their color scheme and pictorial elements evolve in line with the dramaturgy over the course of the magazine article from threatening (soldiers and tanks in dark colors) to informative and clear (the German Chancellor wearing blue in front of the German coat of arms) to inviting and cheerful (birds and bicycles in shades of yellow).

In order to increase the article's interactive and immersive character – and thus the recipients' engagement with the topic – a "sound option" (modality) can be turned on. While scrolling, sounds corresponding to the scenes can be heard at different points in the text, starting with the wailing of sirens at the beginning to the chirping of birds and the ringing of bicycle bells on a street at the ending.

Finally, the aspect of exchange should be briefly mentioned. In journalism, communication is classically unidirectional, with journalists "sending" and recipients "receiving" (with the exception of letters to the editor, which may be published). With the Internet and the emergence of new journalistic, semi-journalistic, and other platforms for content exchange, this division has been softened. In constructive journalism, this means seeing the bidirectional exchange between journalists and recipients as an asset. In particular, contributions (such as the example selected here) that use the SusTelling approach can lead to increased involvement. And this involvement, in turn, can lead to a constructive exchange that reinforces the effects of constructive reporting. These include a better understanding of the topic, an increased willingness to engage further with the issue, and an increased willingness to act.[4]

The sample article generated a lively discussion among *Perspective Daily* members (i.e., its subscribers who are allowed to post comments). This inspired Felix Austen and me the following year to produce a post, using a closely related approach, titled "How far would you go to save yourself?" In it a fictive "green guardian" is elected chancellor, who calls for a strict sustainable social order and ensures compliance through sharp controls (Urner & Austen 2017).

6.3.4 Craft: Transferring Scientific Working Methods to Journalism

The last aspect of constructive journalism will only be briefly outlined here, since although it is fundamental, it does not play a central role in SusTelling. It is about how journalistic reports are written and how a scientific way of thinking and working can be transferred to journalism. In order to give a deeper answer to the what-now question, some proponents of constructive journalism also speak of evidence-based or even replicable journalism. This can take the form of citing sources in references to online articles (by hyperlinks clearly displayed but not too distracting as symbols in the margin) as well as further information. The latter can be shown or hidden by clicking on a symbol in the text. They provide background knowledge, for example, in the form of definitions, legal texts, or more detailed descriptions of research.

In addition to presenting the end product of science, its findings, the creative process also plays an important role in transferring the craft of doing science to the everyday work of doing journalism. This includes, most importantly, a collaborative way of working, which – based on the process of peer review – includes several feedback loops. Seeing rough drafts of a text marked up in color is not something negative, it is constructive criticism that improves the quality of a report – and a crucial learning process for many journalists.

6.4 Constructive Journalism as an Opportunity for the Future

After situating constructive journalism in a larger context in the introduction to this chapter, I would like to conclude with a brief assessment or perspective on its role to date and its possible role in the future. As the title of this chapter reveals, in my view, the constructive approach and SusTelling make a "perfect pair" since sustainability by definition involves a future-oriented perspective that asks the what-now question. When the what-now question is answered collectively, it is not part of a journalistic trend but a fundamental responsibility in a democracy. Which direction do we want to take as a society? And how can we overcome adversity and get a little closer to this goal? In this sense and with this purpose in mind, the working principles of constructive journalism are of overarching importance. Only if journalists begin thinking about and researching the what-now question in their daily work can they report in a way that fulfills journalism's core tasks of helping individuals in a democracy form opinions and make decisions as well as possible. From my point of view, the goal would therefore be that constructive journalism would no longer need to be labeled constructive because all journalism would be constructive. Making constructive journalism a tautology would be a further connection to sustainability, and to the goal that all decisions and actions are aligned with the principles of sustainability.

In the past 5 years, I have had the privilege of giving numerous workshops, seminars, and lectures in a journalistic context, as well as countless discussions with media professionals in a wide variety of positions. On the basis of these experiences, I see the biggest hurdles to achieving the goal I have outlined in the entrenched structures in everyday journalistic life and increasing financial pressure. Only if we as a society ensure that journalism is not financed by special interests such as political parties or companies can we open the path to constructive reporting.

To break up these structures, we need engaged people in leadership positions who are open to the findings of psychology and the cognitive sciences. This turning point begins with the realization that as there is no such thing as objective reporting we need to better understand how recipients process information.

My greatest hope, however, comes from observing that change is afoot in the field of journalism, in Germany and around the world. Anthropogenic (man-made) climate change has propelled it out of its eco-niche and into the mainstream of journalism. Likewise, constructive journalism is no longer a fad or fringe phenomenon that has to fear it will soon be replaced by the next trend. Constructive journalism, especially in narrative form in the context of sustainability, is a way of thinking and working that more and more colleagues are embracing and that is continually undergoing new forms of expression.

Notes

1 The journalist Claas Relotius, who had worked as an editor at *Der Spiegel* news magazine since 2017 and had previously written for other media, was publicly exposed as an impostor in December 2018. Relotius had already attracted attention as a potential swindler

in 2014 when a reader had noticed inconsistencies in his article about a hair salon in the *Neue Zürcher Zeitung* newspaper. In 2017, there were doubts about the correctness of some of Relotius' texts in the weekly newspaper *Die Zeit*, among others. But it was the investigative research of *Spiegel* colleague Juan Moreno, with whom Relotius had worked together on the report "Jägers Grenze", that brought about the fall of the award-winning journalist. (For a summary in English, see the reporting in The Guardian at https://bit.ly/3vwZwHB; see also the final report of the investigation commission at *Der Spiegel* at https://bit.ly/37k9ZsH). Numerous editorial offices increased their efforts to avoid subsequent cases by taking additional measures (see also Moreno 2019).

2 The concept of planetary boundaries was first proposed in 2009 by a group of scientists led by the Swedish sustainability scientist Johan Rockström and the Australian climate scientist Will Steffen (2009).

3 This is shown, for example, by the results of the so-called ignorance test conducted by the Swedish Gapminder Foundation. In 2014, the test was also conducted in Germany in cooperation with the news magazine *Der Spiegel* (Gapminder 2020).

4 These are precisely the relationships that initial (pilot) studies on constructive and solution-oriented journalism show. A constantly updated overview of studies can be found on the website of communication scientist Kayser Lough: http:// www.kyserlough.com/solutionsjournalism.html.

References

Alfter, B. (n.d.). *The Importance of Following Up*. [https://bit.ly/3N6SaB9]. Accessed on May 20, 2022.

Austen, F., & Urner, M. (2016, December 28). *In 5 Jahren ist der Notstand der einzige Weg für Deutschland* [In 5 Years, the State of Emergency Is the Only Way Forward for Germany]. [https://bit.ly/3OPhKf5]. Accessed on May 20, 2022.

Berglez, P., Olausson, U., & Ots, M. (eds.) (2017). *What Is Sustainable Journalism? Integrating the Environmental, Social, and Economic Challenges of Journalism*. Frankfurt: Peter Lang.

CACE – Council Action in the Climate Emergency (n.d.). *History of Climate Emergency Action by Councils*. [https://bit.ly/3t5PGeq]. Accessed on May 20, 2022.

EU-Parliament (2019, November 29). *The European Parliament Declares Climate Emergency*. [https://bit.ly/3F1GJrb]. Accessed on May 20, 2022.

Fredrickson, B. L. (2013). Positive emotions broaden and build. *Advances in Experimental Social Psychology*, *47*, pp. 1–53.

Gapminder (2020). *The Ignorance Project*. [https://bit.ly/3Kzx10n]. Accessed on May 20, 2022.

Guo, L., & Volz, Y. (2021). Toward a new conceptualization of journalistic competency: An analysis of U.S. broadcasting job announcements. *Journalism & Mass Communication Educator*, *76*(1), pp. 91–110.

Knodt, S. (2019, July 26). *Konstanz ruft Klimanotstand aus* [Constance Declares Climate Emergency]. [https://bit.ly/38CWNn2]. Accessed on May 20, 2022.

Kovach, B., & Rosenstiel, T. (2007). *The Elements of Journalism: What Newspeople Should Know and the Public Should Expect*. New York: Three Rivers Press.

LaBar, K., & Cabeza, R. (2006). Cognitive neuroscience of emotional memory. *Nature Reviews Neuroscience*, 7(1), pp. 54–64.

Lough, K., & McIntyre, K. (2021). A systematic review of constructive and solutions journalism research. *Journalism*, 146488492110445.

Lünenborg, M., & Medeiros, D. (2021). Journalism as an affective institution. Emotional labor and the discourse on fraud at Der Spiegel. *Journalism Studies*, *22*(12), pp. 1720–1738.

Maier, S. F., & Seligman, M. E. (1976). Learned helplessness: Theory and evidence. *Journal of Experimental Psychology: General*, *105*(1), pp. 3–46.

McIntyre, K. (2016). What makes "good" news newsworthy? *Communication Research Reports, 33*(3), pp. 223–230.
McIntyre, K., & Gyldensted, C. (2017). Constructive journalism: Applying positive psychology techniques to news production. *The Journal of Media Innovations, 4*(2), pp. 20–34.
McKibben, B. (2016, August 15). *A World at War.* [https://bit.ly/3F6cNKp]. Accessed on May 20, 2022.
McNair, B. (2009). Journalism and democracy. In K. Wahl-Jorgensen & T. Hanitzsch (eds.), *The Handbook of Journalism Studies* (pp. 237–249). London: Routledge.
Moreno, J. (2019). *Tausend Zeilen Lüge. Das System Relotius und der deutsche Journalismus* [A Thousand Lines of Lies. The Relotius System and German Journalism]. Berlin: Rowohlt.
Oxford Dictionaries (2022). *Climate Emergency.* [https://bit.ly/3LxLD1q]. Accessed on May 20, 2022.
Reitan, C. (2016). Gesellschaft im Wandel – Journalismus im Wandel. Der Nachhaltigkeits-Journalismus muss raus aus der grünen Ecke und in die Mitte der Gesellschaft – Ein Plädoyer [Society in transition – Journalism in transition. Sustainability journalism must get out of the green corner and into the middle of society – A plea]. In G. Michelsen & D. Fischer (eds.), *Nachhaltigkeit und Journalismus. Erkenntnisse und Impulse aus Wissenschaft und Praxis* [Sustainability and Journalism. Insights and Impulses From Science and Practice] (pp. 137–146). Bad Homburg: Verlag für Akademische Schriften.
Rockström, J., Steffen, W., Noone, K., Persson, Å., Chapin, F. S., Lambin, E. F., Lenton, T. M., Scheffer, M., Folke, C., Schellnhuber, H. J., Nykvist, B., Wit, C. A. de, Hughes, T., van der Leeuw, S., Rodhe, H., Sörlin, S., Snyder, P. K., Costanza, R., Svedin, U., . . . Foley, J. A. (2009). A safe operating space for humanity. *Nature, 461*(7263), pp. 472–475.
Schudson, M. (2001). The objectivity norm in American journalism. *Journalism, 2*(2), pp. 149–170.
Seligman, M. E. P., & Csikszentmihalyi, M. (2000). Positive psychology: An introduction. *American Psychologist, 55*(1), pp. 5–14.
Sheldon, K. M., & King, L. (2001). Why positive psychology is necessary. *American Psychologist, 56*(3), pp. 216–217.
The Economist (2017, May 6). *The World's Most Valuable Resource Is No Longer Oil, but Data.* [https://econ.st/3LAfwON]. Accessed on May 20, 2022.
Urner, M., & Austen, F. (2017, December 28). *Wie weit würdest du gehen, um dich zu retten?* [How Far Would You Go to Save Yourself?]. [https://bit.ly/3ksM0OO]. Accessed on May 20, 2022.
Urner, M., & Langeslag, H. (2018, March 2). *Why Objective Journalism Is a Fata Morgana.* [https://bit.ly/3GwU1g5]. Accessed on 1 May 2022.
Willnat, L., Weaver, D. H., & Choi, J. (2013). The global journalist in the twenty-first century. *Journalism Practice*, 7(2), pp. 163–183.
Wu, T. (2016). *The Attention Merchants: From the Daily Newspaper to Social Media, How Our Time and Attention Is Harvested and Sold.* London: Atlantic Books.
Zhou, N. (2019, November 21). *Oxford Dictionaries Declares 'Climate Emergency' the Word of 2019.* [https://bit.ly/3y0oIrM]. Accessed on May 20, 2022.

7

LEARNING FROM HUMBOLDT – STORYTELLING AS A METHOD OF SCIENCE COMMUNICATION

Dorothee Menhart and Sabine Hoscislawski

7.1 Introduction

Alexander von Humboldt was a great storyteller. Full of emotion, the explorer and natural philosopher reported on the current state of science in his famous Cosmos lectures in the late 1820s. He brought back carefully rendered and artistically magnificent illustrations of the observations he made during his expeditions. A gifted communicator of his scientific discoveries, he is said to have written nearly 50,000 letters, among which are some truly good stories. Storytelling in science communication is more than 200 years old – and has lost none of its relevance. Whether in exhibitions, science podcasts, lectures, or web videos – whatever the format of science communication, storytelling is a tried and tested means of communication, even though it is repeatedly met with criticism – often from scientists themselves (Bucchi 2014; Görke & Ruhrmann 2003; Rödder 2009; Schäfer 2009).

In this chapter, we present projects from the practice of science communication using the storytelling method with a focus on sustainability communication. As staff members of Science in Dialog (WiD),[1] a German organization for science communication, we focus on successful examples that have been developed, discovered, or awarded prizes by WiD and other actors in science communication in recent years. In the second part, we outline why storytelling in science communication is controversial. In the concluding part, we address questions about the impact of storytelling in science communication.

7.2 Storytelling in Science Communication: Examples From Sustainability Communication

Today, countless formats make use of storytelling or individual elements of this method of presenting and communicating knowledge. Examples include the *Fast*

DOI: 10.4324/9781003326144-11

Forward Science competition (Wissenschaft im Dialog 2020a), which since 2013 has been producing web videos that present topics, ideas, and phenomena from research, and the *MS Wissenschaft*, a ship that tours Germany offering the public science exhibits, each of which follows a distinct plot.

In science slams – modeled on the somewhat older poetry slams – young researchers present their findings in a personal, authentic, and entertaining way to an audience that grows year by year. New formats such as 360° videos attract attention, not only because they tell stories but because they can also show scenarios of a future that might happen – a past–present–future view of, for example, issues in sustainability.

Outstanding formats for communicating sustainability research can always be found in an initiative of the German Ministry of Education and WiD. *Year of Sciences* has included such themes as *Future Project Earth* (2012) and *Seas and Oceans* (2016/17) and the Year of Science 2020/21 *Bioeconomy*. Some examples are briefly presented later. They were selected on the basis of their vividness, the diversity of their format, and their originality.

Jonah and the Deep-Sea Monsters

As part of the Year of Science *Seas and Oceans*, the dance theater video *Jonah and the Deep Sea Monsters* was created and toured 13 northern German cities in the summer of 2017. It tells a classic hero's journey: Together with Jonah, the hero of the play, children from the age of seven go on a daring expedition. Their goal is to find the mysterious tears of the mermaids (which turn out to be plastic waste). The story's main character and hero is portrayed by a dancer. The backdrop is the deep sea brought to life through animated drawings and video projections. On her journey, the hero is accompanied by various deep-sea dwellers who show and explain things to her, encouraging her to think. Supposedly scary creatures like the deep-sea angler fish turn out to be clever nature-based solutions and supposedly beautiful phenomena, the microplastic particles called the "tears of the mermaids" by researchers, turn out to be the real threats. The musical background reflects the atmosphere and dramaturgy of the scenes and frames the narrative with mysterious sounds (tonality).

The story and how it was realized is meant to encourage children and families to actively reflect on the social significance of the seas and oceans while highlighting the crucial importance of the science and research being done on these topics. Marine researchers were invited to selected events and engaged with the audience in a question and answer session, giving background information on what the audience had experienced in the theater.

Explainer Videos in the Year of Science 2020/21

In the 2020/21 Year of Science *Bioeconomy*, the German Ministry of Education together with WiD is communicating the concept of bioeconomy to reach a broader audience. Their homepage identifies four areas covered in the campaign: plant diversity and innovation, soil health and material cycles, microorganisms and technology drivers, and bioeconomy and change (BMBF 2020a). Each of these topics is summarized in a short explainer video lasting a little less than two minutes. The videos and accompanying texts communicate clear values to the recipients that go hand in hand with the normative call for promoting awareness of sustainable development. The goal is to bring about sustainable changes in the use of resources, production processes, and consumer behavior.[2]

The explanatory videos (see, e.g., BMBF 2020b) convey the story behind this Year of Science. As its initiators explain, our current economic practices and resource use must be converted to bio-based products and holistically conceived models to secure our standard of living and enable future generations to have lives worth living (BMBF 2020a). The short videos present complex interrelationships of the bioeconomy in an easily understood and interesting way. Each video presents four facts in more detail. The targeted use of stylistics and tonality in the presentation encourages the recipient to perceive the topics as simple, easy-going, and approachable. The video features upbeat contemporary background music and appropriate imagery (e.g., the video on soil health and material cycles shows insects crawling around in their soil habitat). Facts are announced by a flashing light bulb and a pinging sound to create a moment of recognition.

Earthbook – Planet Earth Has a Social Media Account

During the Year of Science *Future Project Earth*, a prime example of storytelling in sustainability communication was developed. For the German *Sustainability Action Day* in 2012, the Year of Science together with the Climate Media Factory presented *Earthbook* (BMBF 2012), a web video of the Earth online.[3] The video shows a platform (with all the features of Facebook) that powerfully illustrates the effects humanity's actions have had on the Earth since industrialization and the rise of a consumer society – and how only slowly have we discovered our responsibility toward the Earth.

The animated short film uses the language and aesthetics of social network platforms to highlight climate change, environmental degradation, and the resulting consequences. The plot follows a thought experiment: What would planet Earth post on its profile about humanity if it could? In fast forward speed, the Earth develops a virtual relationship with humans – and soon raises the question of whether it wants to be "friends" with the species that is overexploiting its natural resources. The positive message of the film is: Yes, it does. After all, many individuals are working flat out to preserve the Earth. And an important key to this is science and research.

The plot of *Earthbook* follows a linear storyline with a chronological history of evolution showing the development of humankind toward an industrial society with all its environmental and climate threats as well as the attempts to counteract them. This focus on history since the origins of human beings, a vanishingly small period compared to the history of the Earth, emphasizes the conflict between humanity and our environment. The main character of the narrative is the planet Earth, and the plot unfolds with a personalized Earth together with other characters in the form of planets, plants, animals, and then humans. The various characters going about their activities seem very "human". In addition to writing posts and speaking, for example, the Earth shows human emotions such as being happy, sick, and annoyed.

The narrative of *Earthbook* conveys values and a normative statement. The thoughtless treatment of the environment and climate and its impact on the Earth is highlighted to bring about changes in people's actions. Personalizing and humanizing the earth makes the message easy for the audience to understand. The Earth's suffering, expressed in human emotions, can help each recipient to redouble their efforts and rethink whether their actions contribute to sustainability. The positive changes brought about by the Kyoto Protocol and the two-degree target, which the Earthbook user Earth "likes", also point to a course of action that the audience can now relate to their behavior.

An Educational Escape Game: BioEconomyNow!

In the Year of Science 2020/21 Bioeconomy, the House of Science in Braunschweig and WiD created an educational escape game called BioEconomyNow! (bioeconomy-now.de) to promote education about the environment. The game is geared toward children and young people aged 10 and older, young adults, and families and can be played in teams of four to eight people at various locations in Germany during the Year of Science 2020/21 (Wissenschaft im Dialog 2020b). The escape game is made up of various tasks that have to be solved in teams, allowing the participants to learn about the bioeconomy and its advantages in a playful way. They are immersed in a fictitious future world called Horizon and have just under an hour to convert its society into a bioeconomy, including a transformation from a petroleum-based economy to one based on sustainable use of renewable raw materials. In the process, the two teams must solve puzzles, thoroughly investigate the use of space and various objects, and make decisions that directly impact the planet's environment, economy, and population. In this way, they learn about the ideas and content of the bioeconomy through play and engage with social issues. The narrative underlying the game format ensures that a dramaturgical arc of tension is created. The fact that the participants themselves are part of the game and the narrative as characters means that they are particularly challenged to engage with the content. The game only works if they are involved. Interactivity and decision-making options allow the players to determine the narrative within a framework of possibilities, which makes sensory perception more memorable than in a classic game and allows participants to identify more strongly with the content. After the game, a support team helps participants learn more about both the research behind a bioeconomy and how it can be practically implemented. Participants are thus able to reflect on what they have experienced and consolidate their learning experiences. The mobile format of visiting festivals in different regions allows the game to reach people in a low-threshold and spontaneous way. A similar version is also available as an online game. Another example of an escape game in English is "Unlock the Future" (Pater et al. 2020).

A Computer Game ENERGETIKA 2010

The computer game ENERGETIKA 2010[4] was developed for the 2010 Year of Science *Energy* to communicate the complexity of sustainable energy supply. In 2011, it was designed for young people aged 15 and over and adults and won the prize in the serious game category of the German Computer Game Award.[5]

In ENERGETIKA 2010, the goal is to ensure the energy supply of the fictitious country of Energetika until 2050 through a balanced energy mix. Keeping electricity prices at an appropriate level and avoiding power outages are just some of the challenges to be overcome. Help is given by consultants, who provide detailed information on the different forms of energy. The game is won by achieving an energy mix that balances its social, economic, and ecological aspects. The narrative behind the game assumes that a sustainable energy mix is not possible by following economic criteria alone and that ecological impact and social acceptance must be given equal consideration. Details such as the possibility of involving the population in a participative process before new power plants are built make abstract processes more tangible. As in the escape game BioEconomyNow! the player determines how the story will unfold through their strategy and the decisions they make.

ENERGETIKA 2010 sensitizes players to sustainability issues in the field of energy. They become aware of the complexity of energy supply and, to successfully complete the game, they must learn more about the subject by consulting tips on the game and an electricity handbook. This awareness of energy issues is also impressively reinforced by accessing data on the energy mix in Germany. The game requires players intensively study the issue as well as develop a strategic approach to facilitate transfer into long-term memory.

Online Comics: The Adventures of Alex and Bioman

Graphic novels and educational comics, nonfiction, and science comics have been used in science communication for many years now. One of the best-known examples is *The Great Transformation* (Hamann et al. 2014), in which climate researchers become comic book heroes. In nine recorded interviews, researchers from the natural sciences, engineering, political science, economics, and law tell what we know about climate change today. They report on desertification, path to a CO_2-free economy, and outline the consequences of unchecked climate change on the global economy. The experts who have their say were members of the German Advisory Council on Global Change (WBGU) at the time, and the comic strip is based on the WBGU report *World in Transition-Social Contract for a Great Transformation*.

Another example of climate and sustainability communication is the comic strip *The Adventures of Alex and Bioman.*[6] The comic looks at the issues of sustainability, forestry, and bioeconomy and explicitly includes the controversial use of wood. This topic is explored through the adventures of Alex, a dachshund belonging to a forester, and Bioman, an environmental superhero. The narrative follows a linear chronological development and has a clear dramaturgy. The comic begins with the dachshund Alex in the Black Forest, who introduces his tranquil life in the forest and tells the recipient about the forest ecosystem, forestry in Germany, and the uses of wood. Then the story picks up speed. A path cut through the forest leads to the fictive town of Nuvos, where the World Economic Forum is meeting. Its participants are discussing the climate and environmental crises and have unanimously identified the global demand for oil as their cause. Environmentally conscious researchers present the solution to resource scarcity and environmental crisis: Bioman. This is the beginning of the adventures of Alex and Bioman, who then take up the fight against the environmental destruction caused by Mr. Petrolius.

The comic's entertainment value contributes to a good understanding of the complexities of the environmental crisis and the economy that come together in the bioeconomy. This is aided by the personalization of different factors in this conflict. The comic characters stand for a stakeholder group or a figure in the crisis – for example, Alex and Bioman as saviors and heroes fight against the environmental crisis and resource scarcity, Mr. Petrolius as an anti-hero perpetrates and benefits from an environmentally destructive energy system, Martin Mustermann is a typical naive citizen who doesn't want to give up his accustomed, but destructive, lifestyle for fear of change. The arc that is drawn between the tranquil Black Forest of Alex the dachshund and the global economic and political entanglements that Bioman and Mr. Petrolius represent does a good job of showing the interrelationships between different levels in the climate crisis. The comic strip's tone is neither one-sidedly positive nor negative. It addresses both environmentally destructive behavior and sustainability issues, such as when Bioman is captured by Mr. Petrolius and then, of all times, his green power battery fails.

Both *The Great Transformation* and *The Adventures of Alex and Bioman* reveal the complexity of the environmental crisis without dumbing them down. The recipient is implicitly encouraged to become aware of their position within this complex situation and to better understand the narratives of the stakeholder groups involved.

The examples discussed earlier all impressively illustrate that Storytelling for Sustainability (SusTelling) is now an established part of science communication and how it contributes to education for sustainable development. They also show the different ways SusTelling can be used in science communication. The difficulty level of the ENERGETIKA 2010 game is comparatively high and the game is, therefore, suitable for different age and target groups than, for example, the dance theater video about Jonah and the Deep Sea Monsters. What the examples have in common is that they break down complicated issues so they are understandable for their target groups and they lower the threshold for individuals to engage with complex topics. This effect is achieved through interactive elements, the entertainment factor, the opportunity to identify with individual characters, emotional appeal, and the creation of references to everyday life. In addition, all the formats illustrated in these examples make a normative call for recipients to form an opinion about sustainability issues based on scientific facts.

The approach of "entertainment-education" has thus been taken up by science, in part deliberately but possibly also unconsciously. Its proponents argue that combining educational content and entertainment factors, such as watching a television series whose characters address sustainability issues, is a promising communication strategy for learning about sustainable lifestyles.[7] Depending on the example, however, the goal of the format is not solely communicating information about sustainability issues. It is also about attempting to generate a general understanding of the role of research and science, for example, through the opportunity to talk with researchers after performances of the dance theater video *Jonah and the Deep Sea Monsters*.

7.3 Why Is Storytelling Controversial in Science Communication?

From the beginning, the rapid development of storytelling in science communication has been accompanied by discussions about the extent to which scientific knowledge is suitable for being communicated through entertaining and necessarily simplistic narratives. At the Forum Science Communication in 2018, this issue was addressed when Julika Griem, vice president of the German Research Foundation asked in her keynote speech: "Why is it that everyone is so interested in narrative communication?" (2018). This made the hundreds of science communicators who are constantly on the lookout for good stories in their universities and research institutions sit up and take notice. Because Griem, a narrative researcher of all people, had hit a nerve, her criticism: The universal boom in storytelling in science communication, with narrativization and eventization, risks simplifying things in a problematic way.[8] In its place, Griem pleaded that more be "gently asked" of the audience. She described her thesis of simplification with a narrative talent of her own:

> On television, in nature documentaries, we follow courageous researchers on their journeys to the "last frontiers" of biodiversity accompanied by a soft narrative voice and swelling orchestral music. . . . And medical expertise is brought to us in a television series like "In Friendship".

Children's universities, chemistry olympiads, and math competitions also testify to a "culture of spectacle". In her analysis "science communication is taking on forms that are increasingly competitive" and is often about producing events that are entertaining and immediately captivating, that have an immediate appeal to the senses, are emotionalizing, and not in the least argumentatively complex or overly demanding. The audience's needs and interests are catered to as if "science communication was all about catching a Flixbus to the next science slam".

The discussion among science communicators at the conference then developed as follows: Isn't this catering to the needs and interests of the audience sometimes unavoidable? Isn't it sometimes even necessary? When populists are spreading fake news at the same time, can science communication do without it? Isn't it important to first awaken an interest in research and science, so that the fundamental value of science as part of a democratic society becomes apparent even to those who perhaps don't like others to "gently ask" slightly more of them? Isn't understanding the audience perhaps one of the very basic tasks of communication, so that, in a second step, the actual goal of science communication can be achieved, namely, to empower as many people as possible to arrive at a critical understanding of scientific facts and to make informed decisions?

This dual responsibility – on the one hand, to involve the public in scientific issues so that they do not simply dismiss them out of hand and, on the other, to maintain standards in communicating the sometimes dry workings and logic of science – poses major challenges for science communication. Griem advocates showing science as it is, not "as an adventurous hunt for the next big thing" but "as an ongoing search that is rather unspectacular in many respects". Science is not only about breakthroughs but also about perseverance.

It is not least against this backdrop that WiD continually encourages scientists to focus not only on spectacular research results when communicating their work but instead on the methods and cognitive processes of their research. This would portray a more realistic picture of science. At the same time, clear communication about the research process would also demonstrate the value of scientific knowledge and set it apart from other forms of knowing – invaluable in times when fake news is becoming increasingly common.

Controversies about popularization and eventization in storytelling exist not only because – explicitly desired by some, criticized by others – they reduce complexity. Above all, there is little evidence of the extent to which narration, emotionalization, and other features of storytelling affect recipients. Do they become more engaged with the content thanks to well-told stories about science? Does their view of certain facts or their attitudes toward science in general change? The evidence is scanty.

WiD, therefore, launched a research project (2019–2023) on science communication to determine the effects of different formats of science communication. What elements, the *Impact Unit – Impact and Evaluation in Science Communication* research project asks, must a project contain to affect cognitive attitudes such as interest and subjective awareness? What makes a format more likely to affect evaluative attitudes or behavioral intentions? Questions like these have been investigated for some formats in science communication at a case level. However, they have not yet been systematically compared and evaluated. Certainly, this research project of WiD will also produce some new insights into the effect of narration and emotionalization as well as methods of storytelling in sustainability communication.

Notes

1 Wissenschaft im Dialog is the organization for science communication in Germany. WiD brings science and the public together in a conversation, promotes awareness of the social significance of science, and strengthens understanding of its processes and findings. To this end, WiD organizes discussions, school projects, exhibitions, and competitions throughout Germany and operates online portals on science and science communication. WiD is constantly developing new communication formats that strengthen the dialogue with society, bring controversial topics into focus, and reach new target groups. The non-profit organization was founded in 2000 by the major German science organizations on the initiative of the Stifterverband. Foundations were added as partners. WiD receives substantial support from the German Federal Ministry of Education and Research.
2 It should be critically noted here that the explanatory videos, due to their brevity and exclusive focus on the benefits of the bioeconomy, disregard the controversial discussion of some areas of the bioeconomy. They therefore serve primarily as an introduction to the topic.
3 Wissenschaftsjahr – the Year of Science – has uploaded a copy to YouTube: https://bit.ly/3PGJ281.
4 It was developed by the non-profit institute Dialogik together with the University of Stuttgart, the Jülich Research Center (KFA), the German Aerospace Center (DLR), and other scientific institutions (BMBF 2010). The game can unfortunately no longer be played with current browsers (https://bit.ly/3kuVJnS).
5 Examples in English can be found in the projects listed by the Playing For The Planet initiative (playing4theplanet.org). Among the features projects is the Green Game Jam, a competition between game studios to find meaningful ways to engage players with climate change.
6 The idea and realization of the comic strip come from Dr. Alex Giurca and Dr. Markus Hebener of the University of Freiburg, one of the winning teams in WiD's university competition in the Year of Science Bioeconomy. It can be accessed at alex-bioman.de.
7 More detailed explanations of the theory and communication strategy underlying entertainment education, its potential and research approaches, especially in mass media, can be found in Reinermann et al. (2014).
8 Griem is not alone in this criticism. A similar point had been made by Michael Dahlstrom and Marty Kaplan in their talk "Science of Storytelling; Storytelling of Science" at the Arthur M. Sackler Colloquium "Advancing the Science and Practice of Science Communication: Misinformation About Science in the Public Sphere" held in Irvine, CA on April 3–4, 2019 (a video recording is available at: https://bit.ly/3kwSexc).

References

BMBF – Bundesministerium für Bildung und Forschung (2010). *Wissenschaftsjahr 2010 – Energie. Energetika 2010* [Science Year 2010 – Energy]. [https://bit.ly/3kvTkcx]. Accessed on May 20, 2022.

BMBF – Bundesministerium für Bildung und Forschung (2012). *Wissenschaftsjahr 2012 – Erde* [Science Year 2010 – Earth]. [https://bit.ly/3s2ZVPU]. Accessed on May 20, 2022.

BMBF – Bundesministerium für Bildung und Forschung (2020a). *Wissenschaftsjahr 2020|21 – Bioökonomie* [Science Year 2020|21 – Bioeconomy]. [https://bit.ly/3vvuzDO]. Accessed on May 20, 2022.

BMBF – Bundesministerium für Bildung und Forschung (2020b). *Wissenschaftsjahr 2020|21 – Bioökonomie. Pflanzenvielfalt und Innovationen* [Science Year 2020|21 – Bioeconomy. Plant Diversity and Innovations]. [https://bit.ly/3kt5IKn]. Accessed on May 20, 2022.

Bucchi, M. (2014). *Science and the Media: Alternative Routes to Scientific Communications.* Routledge studies in science, technology and society. London: Routledge.

Görke, A., & Ruhrmann, G. (2003). Public communication between facts and fictions: On the construction of genetic risk. *Public Understanding of Science*, *12*(3), pp. 229–241.

Griem, J. (2018). *Forum Wissenschaftskommunikation 2018. Keynote. Zumutungen. Wissenschaftskommunikation und ihre Widersprüche* [Forum Science Communication 2018. Keynote. Impositions. Science Communication and Its Contradictions]. [https://bit.ly/3Fcfpqb]. Accessed on May 20, 2022.

Hamann, A., Zea-Schmidt, C., Leinfelder, R. R., & Hartmann, J. (2014). *The Great Transformation: Climate – Can We Beat the Heat?* Berlin: Verlagshaus Jacoby & Stuart.

Pater, E., Harms, C., & Smit, K. (2020). *Unlock the Future – DIY Escape Game*. [https://bit.ly/3vvq8c8]. Accessed on May 20, 2022.

Reinermann, J. L., Lubjuhn, S., Bouman, M., & Singhal, A. (2014). Entertainment-education: Storytelling for the greater, greener good. *International Journal of Sustainable Development*, *17*(2), pp. 176–191.

Rödder, S. (2009). Reassessing the concept of a medialization of science: A story from the "book of life". *Public Understanding of Science*, *18*(4), pp. 452–463.

Schäfer, M. S. (2009). From public understanding to public engagement. *Science Communication*, *30*(4), pp. 475–505.

Wissenschaft im Dialog (2020a). *Fast Forward Science*. [https://bit.ly/38CVtR5]. Accessed on May 20, 2022.

Wissenschaft im Dialog (2020b). *BioEconomy Now!* [https://bit.ly/3Ly9Uob]. Accessed on May 20, 2022.

8

STORYTELLING IN CORPORATE COMMUNICATIONS – WHEN, WHY, AND HOW STORYTELLING WORKS IN PERSONNEL TRAINING

Jennifer Fritz

This chapter summarizes why storytelling is an important method for knowledge transfer and communication in companies, how it can be used in e-learning, and what needs to be considered when designing digital story training for employees in companies.

For several years, I have been working as a learning experience designer for e-learning with a focus on story-driven training. During this time, I was able to develop a method on how best to develop storytelling training for clients. The experiences I made in the process of designing a story-based e-learning module are the focus of this chapter.

8.1 Storytelling in E-Learning

People have always told each other stories. Even today, we read our children fairy tales and other stories to put them to sleep. We consume stories as movies, series, books, and comics or tell them to each other. Our lives are an accumulation of stories. Evidence suggests that up to two-thirds of our daily conversations consist of stories (Hsu 2008). A large proportion of these stories consciously or unconsciously convey knowledge. The role of storytelling as a learning and teaching tool is now particularly evident in digital formats, such as e-learning. The award-winning project, *Climate Under Pressure* (climateunderpressure.com) by TV5 Quebec and TV5 Monde, is a prime example here. The interactive documentary experience was published in 2016 and demonstrated the sweeping potential of stories to bring, in six episodes, a complex, intangible topic such as climate change into people's daily lives. Its impact on each individual was assessed by a so-called climate barometer that measured the consequences of each decision made by a participating player. The development of story-based e-learning for the wider society

DOI: 10.4324/9781003326144-12

also had an impact on the field of personnel training and found its way into the portfolio of major providers of such training. Even with difficult topics – as the positive feedback on examples in the world of e-learning shows – the use of comics, illustrations, and a good story achieves higher engagement with target groups (Imc 2022).

8.1.1 How and Why Storytelling Works in Learning and Teaching: Facts and Figures

Why is storytelling particularly suitable for the field of learning and teaching? Why are people so open to storytelling in the first place? What motivates them to tell stories and to read, listen, or watch when one is being told? To answer these questions, it is worth taking a closer look at recent storytelling research than was already outlined in this volume (see Chapter 1). Behavioral and psychological research into storytelling has produced some exciting findings here.

Numerous insightful studies and theories in recent decades have come from the fields of brain research and psychology. Roger C. Schank (1990), in his book, *Tell Me a Story: Narrative and Intelligence*, deftly summarizes what links intelligence, learning, and storytelling. He defines intelligence as the ability to solve problems. One must be able to draw conclusions from what has already happened to successfully master the future. To do this, it is important to recall the right experience at the right time and draw conclusions about the current situation based on that experience. The key finding of his research for our purposes here is that the human brain stores experience in narrative logic. We have scripts in our heads that help us understand the behavior of others – if everyone has similar scripts in their heads. Many of our intuitive actions are based on such scripts and much of our early life consists of learning these scripts. So, we have mini-stories in our heads that help us learn rules and thus navigate the world.

John Weich (2014, p. 13) summarizes the findings of decades of research on the definition and meaning of stories as follows: Storytelling is a method of connecting with others. It takes advantage of the fact that humanity is familiar with basic plot structures and archetypes. No matter a person's rank or social standing, the basic scripts are instantly recognizable to a majority of people.

In addition to storing information and sense-making, brain chemistry plays a major role in human devotion to storytelling (Neray 2018; Zak 2014). When we engage with a story, our brain thinks we are living through it ourselves and emits various hormones, the most important of which are cortisol, dopamine, and oxytocin. Cortisol is released when we connect with the hero and share in their excitement and problems. And as a reward for this accomplishment, our brain releases dopamine – the happiness hormone. The third hormone connects us to protagonists and society through empathy. Oxytocin is the hormone responsible for our feeling of solidarity and a sense of community. Researchers call us *homo emoticus* – humans who want to feel included and therefore love storytelling above all else (Weich 2014, p. 22). In a particularly revealing study, Paul Zak (2014) and his research team show that stories influence brain chemistry, which in turn can change people's behavior. Individuals who produced particularly high levels of oxytocin and cortisol during the reception of a story show an increased willingness to donate.

But what does all this have to do with learning in general and e-learning in particular? As mentioned earlier, our brain stores memories in a narrative structure. Cognitive scientist Jerome Bruner (1986) shows that we can remember facts that are parts of stories up to 22 times better than when they are presented alone. This also means that the brain prefers to have knowledge presented in the form it uses to store it – as a story. Moreover, a story follows a certain structure. It describes the causal relationship between events that take place in a certain temporal order and that affect certain characters (Dahlstrom 2014). Haven Kendall (2007) defines a narrative as "a detailed, character-based account of a character's efforts to overcome obstacles and achieve an important goal". In the definition of Storytelling for Sustainability (SusTelling), these are the characteristics of personalization, plot, chronology, and dramaturgy (see Chapter 1).

Narratives therefore not only show relationships but also that actions have consequences. Learners thus become aware – through the characters in a story – of the consequences of their actions. This is done using a format that is treated by our brains as if we had experienced these events ourselves (Paul 2012). Thus, the brain does not distinguish between being told a story or experiencing it ourselves. Storytelling has the potential then to be an effective learning and teaching method. The challenge is to identify when it can best be used and how to deliver a story in a way that learners enjoy working through the learning tasks that are part of it.

8.1.2 On the Role of Fiction in Story Training in E-Learning

Let's start here. What actually defines e-learning? Following an inclusive definition proposed by Sangrà et al. (2012), e-learning can be understood as all forms of teaching and learning that use electronic and/or digital media as "tools for improving access to training, communication, and interaction and that facilitates the adoption of new ways of understanding and developing learning" (p. 152). Typical for storytelling in e-learning is the characteristic of modality (see Chapter 1), that is, the interaction of learners with media content. For example, learners are presented with the task of solving problems and must make choices that determine the further course of the story.

Fundamentally, storytelling for e-learning differs from fictional formats in the degree of freedom given to the designers. In contrast to purely fictional storytelling, in a story-driven training content must be conveyed in a way that is appropriate to the corporate context and the target group. The dramaturgy and plot are thus wrapped around this content as narrative techniques but are only a supplement to the actual content. Educational entertainment is very demanding in terms of craftsmanship since the problem-centered setting of story-driven training in e-learning is shaped by certain roles and expectations that are fundamentally different from a nonwork setting with an entertainment function. What's more, the content is often not typical for classic scripts and its differences must be explained.

In my experience, the attitude of the recipients in story-driven training in e-learning is different from that in fictional formats. Most e-learning in the corporate environment is mandatory. This means, for the most part, that many learners enter this situation with a less than open attitude and thus follow instructions to engage with storytelling content. Team-building and occupational health and safety measures or the teaching of corporate values and norms reflect social sustainability goals in a broader sense – for example, when it comes to compliance measures in companies. With story training in e-learning, such goals or changes are implemented not by imposing rules but by strengthening employees' empathy and reflection abilities.

There are various ways of presenting the structure of a story in e-learning. One of the best known is certainly *Freytag's pyramid* (Bunting 2020; see Figure 8.1).

The structure of story-driven e-learning, in my experience, takes the following form. The exposition introduces the place, time, and major characters. In this phase, the hero is assigned a task, that is, a role in the story. In the second act, the action rises and the conflict intensifies. The first obstacles and resistance appear, making the learner actively solve problems. In the third act, a turning point or climax appears. This is the moment when things become better or worse. In the fourth act, falling action, there is a showdown between the protagonists and their antagonists. Here, it becomes clear who will be successful in the end. If the learner does everything right, they will carry victory off the pitch as the hero. And the resolution at the end of the story shows exactly that: Was the learner successful?

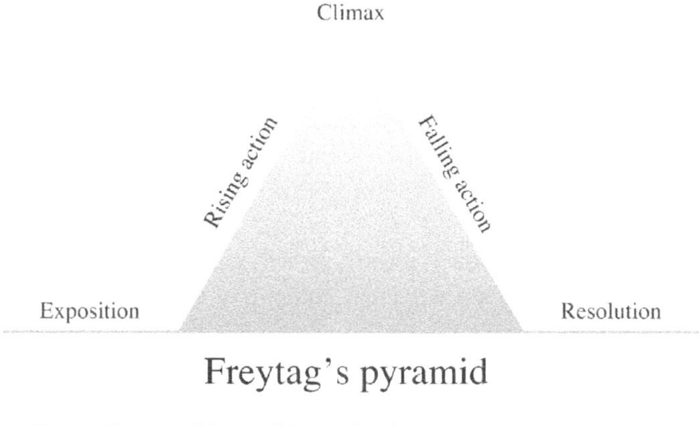

FIGURE 8.1 Freytag's pyramid, graphic: author's own

This classic structure in five acts, which *Freytag's pyramid* depicts, is something many of us have become familiar with in our lessons at school. In the e-learnings I have designed, all of the learning content, and thus most of the story, takes place in acts two to four. The familiar structure makes it easier for the brain to absorb the story together with the content, but this does not guarantee that the training will be successful. Once the learner has been "hooked", it must be maintained through further steps.

We have already seen that hormones, as well as emotions, play a major role in the fact that we humans like to tell and be told stories so much. This means that in story-driven training we need to forge an emotional connection between the plot and characters (the learning content) and the recipients.

Story-driven training is a subcategory in the larger field of e-learning, where the learning and teaching process is supported by various digital tools. Digital media and interactions are used to convey the learning content. In digital story training, e-learning with a story focus, the method of storytelling is also used to make the learning content attractive to the learners.

Christina Maria Schollerer (2017a, 2017b) explains how this can be done using Robert McKee's "hook". A hook triggers emotions that catch the attention of the recipients; it draws learners into the story training and is the reason why they want to see it through until the end. However, a hook alone is not enough to hold attention – obstacles must be overcome along the way, which is why the hook must be followed by a "hold". The hold is what secures our attention for the longer term, an exciting or emotional development in the story. At the end of the story, there needs to be a reward for the effort the story has put us through, the "payoff".

This foundation for story training in my professional practice is summarized in Figure 8.2.

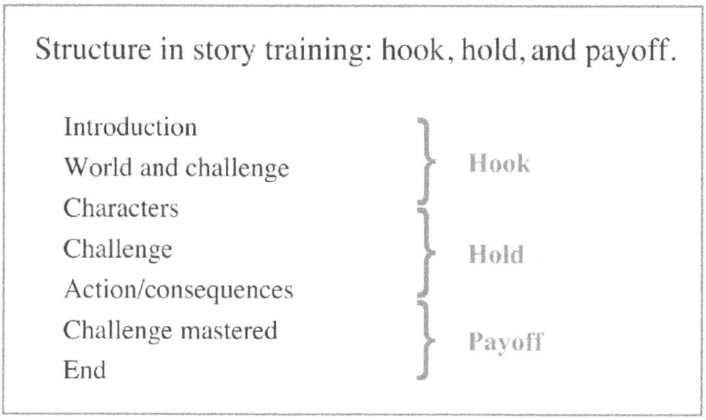

FIGURE 8.2 Hook, hold, and payoff applied to the structure of story-driven training, graphic: author's own

Hook, hold, and payoff illustrated in an example:

Hook	We are in the company Supermarketing. The new employee has their first day and has to overcome some challenges. Business at the agency is fast-paced and full of surprises – and traps.
Hold	Supervisors and co-workers help the trainee. A first challenge has to be mastered – a client or customer wants an order done today and only the new employee has time. Depending on how the learner responds to the task, the agency will have a satisfied or angry customer.
Payoff	After some initial difficulties, the learner masters the situation to everyone's satisfaction. In the end, they receive a badge recognizing their good work and high praise from supervisors and co-workers.

In story-driven training, the hold part can be repeated as many times as needed until the entire learning content has been conveyed.

8.1.3 Application Cases for Storytelling in E-Learning

From my practical experience, there are some situations and factors in which the storytelling method is particularly well suited for the e-learning field:

1. *The training program is optional, but the topic is important. Employees should be "attracted" by a good story*: The number of mandatory training is overwhelming in some industries and companies. Nevertheless, again and again, a topic is identified that is important for the employees. This is where storytelling and the power of word of mouth come into play. If the training program is recommended to others because it is something new and exciting, there can be a snowball effect.
2. *The training program has a very boring, abstract, or difficult-to-understand topic*: Especially among the mandatory topics, and in particular the area of compliance, training content is often difficult to grasp but must be completed every year. Storytelling is a suitable method to make such topics more interesting. When it is a question of changing behavior or values, training can also be hard to follow, especially when it is important to give a concrete example of how such situations can be handled in the future. Packaging different behaviors and values into short stories that learners experience as if they were real can make the learning objectives much more tangible.
3. *The employees do not want to admit that they have deficits on a topic*: There are topics in the company where employees have a "blind spot". Or, less often, it is a topic that employees do not want to admit they need to learn something about. In still other cases, an attempt has already been made to train employees

on a topic, but it has failed and now employees have a negative attitude toward the topic. Storytelling can raise topics to a different level so that employees don't identify its context, paving the way for acceptance and new insights.

In a nutshell, storytelling is particularly suitable for a company when learning content can be subjectively experienced by employees, when it needs to be made more exciting, or when learners need to be convinced to even engage with a topic at all. It is entertainment and education in one – versatile and varied. Of course, storytelling in e-learning is also suitable for other situations and other content. This approach, as explained earlier, not only conveys content but also connects more deeply with learners and helps them overcome situational obstacles.

8.2 What Can Story Training in E-Learning Look Like in Practice?

Let's look at how an abstract topic can be turned into story training using two examples, one from the field of awareness training and the other from the field of change management. First, let's take a brief look at what characterizes these training categories.

Awareness training, or rather behavioral training, deal with topics that can become hazards for the company in daily work. The goal of awareness training is to sensitize employees to adapt their behavior by learning about situational factors and the consequences of their actions. This type of training is usually conducted to prevent a company from having legal and financial problems or loss of reputation. In e-learning, such situations can best be conveyed through story training, since the consequences of actions are an integral part of the story. The protagonist – in many cases the employees themselves – learns by overcoming obstacles and having to live with the consequences of their negligence.

Change management training teaches the process, techniques, and tools that can explain changes to employees in an organization to achieve a planned goal (Biech 2016). The focus is typically on processes, structures, and job roles. Since change takes time in an organization, the employees who will be instrumental in bringing about the change must be trained. Storytelling in e-learning serves to motivate employees to change and creates understanding for the necessary steps. It lets the employees experience the future, showing improvements and justifying why the scenario presented in the story is the right way forward. The following examples make clear how this can be done.

8.2.1 Example 1: Fire Safety Awareness Training as E-Learning in a Company

A company with several thousand employees and several offices worldwide has developed new fire safety regulations. To make the entire workforce aware of these regulations and prevent fires from occurring in office buildings, this company has decided to develop a 30-minute fire safety training and make it mandatory for all employees. Since there is a lot of mandatory training in this company, they are not very popular among the workforce. But supervisors and team leaders need to learn in detail about the risks and, if necessary, alert their teams to potential violations. The learning objectives are that the managers will understand the five areas being trained, recognize violations, and be able to provide appropriate guidance and information to their employees.

The fire safety training includes the following areas:

- Keeping escape and rescue routes clear
- Fire doors
- Open flames
- Personal electrical devices, charging cables, and lithium batteries
- Fire safety in the workplace.

As already mentioned, in contrast to fictional storytelling, the challenge with story training is to convey specific learning content and achieve learning objectives – in other words, the story must fit the training topics. I usually propose three to five ideas for a story-based e-learning program to the client. Each proposal in the following outlines a plot and what perspective the learners will take.

- *Proposal 1*: A threat has been received by the company. An unknown person wants to harm the company by setting a fire. You (the learner), together with the fire safety officer, will try to eliminate any risks. You have little time to find and eliminate all the risks in the office building, inspect the fire doors and check the escape routes. Together with the fire safety officer, you try to avert the threat to your company.
- *Proposal 2*: The worst has already happened. There was a fire in one of the office buildings and some employees were injured. The property damage is immense. But you have a second chance. You can travel back in time and eliminate all the fire safety risks. Save the day, and be a fire safety hero.
- *Proposal 3*: The company has chosen you as the new fire safety officer. You must go through some training and testing to prove you're up to the job. Walk through the building pointing out potential risks to the employees. They don't make it easy for you. Fire hazards are everywhere.

The client company chose the first proposal because it seemed to have the most potential. In the next step, an instructional designer divides the learning content into sections and plans a story. With 30 minutes of learning time and five content chunks, it is a good idea to break it up into about five minutes per topic. A few minutes are needed to introduce the story and some for a well-rounded conclusion. Otherwise, hook and payoff will come up short.

After an idea for the story has been found, the story needs a setting (see Chapter 1). The questions that need to be answered to develop a rough concept for world-building are the following:

- Real setting or fictive setting?
- Future, present, or past?
- What does this setting look like? Which characters or creatures inhabit it?

World-building, as done in a series or computer game, involves finding answers to countless other questions. However, these questions are enough to create a basis for story training in this situation. The questions could be answered as follows: Training takes place in a realistic setting in the present, with a focus on people. The setting is rather urban and takes place mainly in a company building, on a train, and in the home office of one or more of the employees.

Next, we move on to the specific characters playing a role in the setting. This plot requires at least two other characters besides the first-person perspective taken by the learners. The first character works for the company and takes the role of a mentor:

- *Maria, 45, fire safety officer.* She likes to wear jeans and plaid blouses, is nice and approachable, but is proactive when it comes to fire safety. She's been a fire safety officer forever and knows all the regulations by heart. She has great faith that together you can avert the danger in time.

The second character is the antagonist who wants to set the building on fire.

- *Leon, 22, former employee.* He's furious at the company. He prefers to wear black because black always makes you look cool. A year ago he was fired from his apprenticeship. And all because he (in his opinion) made a few small mistakes: taking a few bills from the cash register here, telling his buddies how to get into the building to steal screens and computers there. Too bad he got caught. Now he wants revenge.

This learning content can now be wrapped in a coherent plot. An overview of the story is in the following table:

Storyline	*Learning content*
1. 10 a.m., company headquarters, fire safety officer's office. There is a letter on the desk. It is written with letters cut out from a newspaper and contains the following threat: "I'll burn you down! And no one will be able to prove it. I'll make it look like an accident. Revenge is sweet!" Next to the letter is a business card from the fire safety officer.	Clicking on the hotspot of the business card shows the contact information of the company's real fire safety officer.

(*Continued*)

(Continued)

Storyline	*Learning content*
2. The fire safety officer turns to you and says. "Glad you're here. I need help if we're going to eliminate all the fire risks in the building. It's best if you come with me. We're looking for three kinds of fire hazards".	In a conversation (allowing the learner to influence the plot), the employee learns about their role and the basic categories of fire hazards in a company: • Fire doors and escape routes • Fire safety in the workplace • Electrical devices and lithium batteries
3. Together you set out to check the fire doors and escape routes. The clock is ticking, so the fire safety officer gives you the most important facts on the subject along the way.	In another conversation (allowing the learner to influence the plot), the employee learns quick facts about escape routes, fire doors, and more.
4. You see a fire door in a hallway. You stop and tell the fire safety officer what you notice.	The learner applies directly what they have learned about fire doors and escape routes. They have to find hidden or unseen violations of the fire safety rules.

Elements from gamification, such as points and badges, are used to motivate the learner. In this training program, the following elements emerge from the plot: A clock that runs down to the planned attack on the company makes perfect sense in my view. It reinforces the urgency that was communicated at the beginning of the training session and was built up in the plot. In addition, you could have a map sketched out so the team could gradually work its way through the locations and mark which areas are secured as they move through the building.

These notes are already a rough concept for story training. To develop an adequate story, it is essential to be in close contact with people who are familiar with the context (in this case, the company). Queries to the client about specific processes and scenarios in their company can be made to ensure that the story really fits. This is the only way to obtain information about the context and the topic that will enable the training participants to relate authentically to the story.

8.2.2 *Change Training as E-Learning for New Leadership Requirements*

The second example is about managers learning to represent corporate values. In practice, leadership values are often difficult for employees to understand and implement. Therefore, managers have been asked to practice value-based leadership in a practical scenario. Whether the story is a realistic one is not important to the client. It should be fun and make values something easily grasped. Possible leadership values might include the following:

- Empathy is what separates those who work well with others from those who cannot.
- Business success is never an individual achievement; it is important to recognize the role of the team.
- Everyone makes mistakes, even a leader. Open communication helps everyone avoid repeating that mistake.

The goal is for the learners to live these corporate values in their daily work as leaders. So, in this case, the learners need to be exposed to situations where they must represent values in their actions.

Now there are different possibilities for plots – some more realistic, some more fanciful. If the client is open to something more unconventional, I would suggest the following plot:

Plot

> You (the learner) are leading a team of archaeologists on a quest to find the lost treasure of El Catar. This time everyone is sure they've found the entrance to the sunken temple. And you have found El Catar! But no sooner have you all entered the temple than, to everyone's horror, the door slams shut behind you. Everyone looks to you, the excavation leader. It's up to you to lead the team safely out of the temple, but you will need to delegate the right tasks to the right people at the right time.

Setting

> The training is set in a world similar to the real one - like that of Indiana Jones. There are fantastic temples with adventurous quests for the team to The training is set in a world similar to the real one - like that of Indiana Jones. There are fantastic temples with adventurous quests for the team to solve. There are only humans in this scenario and no imaginary creatures. The culture of this lost civilization is purely fictitious. We know only that they used a hieroglyphic language. Since this story is more about teamwork, technology is at a level that all employees are familiar with. Needless to say, there are adventurous tasks for the team to solve together.

Characters

Besides the learner, there are four other team members:

- *Anna*, linguist, 33 years old. She's an expert in the extinct hieroglyphic language of the Catarians. If you don't know what's written there, she can figure it out.
- *David*, 50 years old. He's an excavation veteran. The MacGyver on the team. Give him a pen and some gum and he'll build something guaranteed to help.
- *Nikolas*, 42 years old. He's a historian. He's not up for hard physical labor, but no one knows more about Catarian culture than he does.
- *Chris*, 26 years old. She's a survival expert and first aider. When it comes to jungle expertise, she's the best.

8.3 Tips for Planning and Designing Story Training in E-Learning

Some tips and hints for planning and designing training where a story plays a prominent role:

> Finding the right plot for a target group in the company or outside it requires knowledge about them. Relevant information can be which age groups are involved – and so which stories the learners grew up with or which narratives are particularly well known in this generation, how many technical tools and gadgets the target group can handle, etc. At the beginning of a project, reflect on what you know about the target group. In the best case, before working on the final concept show the target group the story outline to make sure that they do not immediately reject the story.

Not every topic fits every plot. As explained earlier, there are well-known scripts in every society that a target group will feel particularly comfortable with (see Chapter 8.1.1). Some scripts are universally valid (Booker 2019).

Common story plots that can also be used in business include:

- *The quest*: The hero is torn from familiar surroundings, goes on an (inner) journey, and grows from their suffering until at the story's end they return victorious and transformed.
- *Overcoming the monster*: The main character faces an overpowering opponent, a monster that must be defeated because the hero's world depends on it.
- *From rags to riches*: the narrative of Cinderella is the American dream. The hero works her way up from the poorest of circumstances to the highest of heights.
- *Coming of age*: Perseverance and ingenuity allow the hero in this script to solve every challenge or puzzle. Here, too, it's all about proving oneself and growing as a person.
- *Rebirth*: The theme of this plot is transformation, from an ugly duckling to a beautiful swan.
- *Voyage and return*: Very close to the hero's quest, the hero goes on a journey to discover themselves or for the sake of adventure.
- *Comedy*: Comedy thrives on humor and happy endings. All of the other plots described here can just as easily fail.
- *Tragedy*: Just as we know comedy will end well, we know tragedy will end badly.

Still, not every one of these plots fits every topic. The most challenging part of designing a story training program is figuring out which story fits both the topic and the company.

It is advisable to first do a rough draft of a storyline and then present it to all the departments involved in the training exercise. Espionage and hacker stories in particular can lead to problems with other internal departments. But characters and how they are portrayed (physically and otherwise) can also cause acceptance problems later. In one case, the skirt length of a finished character was objected to and had to be exchanged for work trousers. When you are first conceptualizing a story, it is better to ask too many rather than too few people for their opinion. Without this acceptance, the plot may have to be rewritten – in the middle of its conceptualization or even after completion.

Story training thrives on challenges and antagonistic characters. A classic drama needs conflicts, developments, and solutions. In many companies, however, there is still the attitude that there aren't any problems at all. In this case, it can help to have an antagonist from outside the company throw a spanner in the works. A story needs antagonistic characters and challenges – otherwise, it's not a real story. So, if a client wants to eliminate all conflict from the online story training concept, you will need to explain why this is not a good idea.

In conclusion, training should generally be empowering. We want employees to be able to do more after they have completed a training program than they were able to before it and to draw confidence from this for their work. At the same time, we try to protect employees from possible dangers and precarious situations. Story training of this kind is thus closely related to social sustainability goals. Storytelling in e-learning can help to make supposedly dry, demanding, or abstract topics interesting and attractive, to make our everyday lives – especially at work – safer and tailor it better to the needs of the people working there, as well as connect employees' yet untapped potential with corporate sustainability efforts. Storytelling can use plots, characters, or patchwork narratives to communicate topics important for the future with a high entertainment value.

References

Biech, E. (2016). *Change Management Training*. ATD workshop series. Alexandria: ATD Press.

Booker, C. (2019). *The Seven Basic Plots: Why We Tell Stories*. London: Bloomsbury Continuum.

Bruner, J. (1986). *Actual Minds, Possible Worlds*. Cambridge: Harvard University Press.

Bunting, J. (2020). *Freytag's Pyramid: Definition, Examples, and How to Use This Dramatic Structure in Your Writing*. [https://bit.ly/3MGB6kB]. Accessed on May 20, 2022.

Dahlstrom, M. F. (2014). Using narratives and storytelling to communicate science with nonexpert audiences. *Proceedings of the National Academy of Sciences of the United States of America, 111*(Suppl. 4), pp. 13614–13620.

Hsu, J. (2008). *The Secrets of Storytelling: Why We Love a Good Yarn* [https://bit.ly/3y5GjOU], Accessed on May 20, 2022.

Imc information multimedia communications AG (2022). *Case Study: Compliance Training im Sin City-Stil bei Audi AG* [Case Study: Compliance Training Sin City-style at Audi AG]. [https://bit.ly/3vR5GkS]. Accessed on May 20, 2022.

Kendall, H. (2007). *Story Proof: The Science Behind the Startling Power Of Story*. Devon: Greenwood Publishing Group.

Neray, M. (2018, January 12). *Your Brain on Storytelling*. [https://bit.ly/380BFHn]. Accessed on May 20, 2022.

Paul, A. M. (2012). *Your Brain on Fiction*. [https://nyti.ms/3y5PcYJ]. Accessed on May 20, 2022.

Sangrà, A., Vlachopoulos, D., & Cabrera, N. (2012). Building an inclusive definition of e-learning: An approach to the conceptual framework. *International Review of Research in Open and Distance Learning, 13*(2), pp. 145–159.

Schank, R. C. (1990). *Tell Me a Story: Narrative and Intelligence*. Evanston: Northwestern University Press.

Schollerer, C. M. (2017a, January 20). *Hook, Hold, Payoff – die Magie des Dreiklangs* [Hook, Hold, Payoff – the Magic of the Triad]. [https://bit.ly/38Ls4o4]. Accessed on May 20, 2022.

Schollerer, C. M. (2017b). *Transmedia Storytelling*. Workshop-Lab gehalten auf der Story: First, München, Deutschland, 12 June 2017 [Transmedia Storytelling. Workshop-Lab held at Story: First, Munich, Germany, June 12].

Weich, J. (2014). *Storytelling on Steroids: 10 Stories That Hijacked the Pop Culture Conversation*. Amsterdam: Bis Publishers.

Zak, P. J. (2014, October 28). *Why Your Brain Loves Good Storytelling*. [https://bit.ly/3LzqoMV]. Accessed on May 20, 2022.

Epilogue

STORYTELLING FOR SUSTAINABILITY

A Preliminary Conclusion

Daniel Fischer, Sonja Fücker, Hanna Selm, and Anna Sundermann

Looking back is where things are born.

Dirk von Lowtzow

If, at the end of our volume on storytelling in sustainability communication, we now want to review the individual chapters and draw a conclusion relating to our guiding questions, the question inevitably arises as to which interpretive framework we choose for this: a self-reflective story of discovery, of slow maturation, or even a riddle (Kent 2015). In the end, to what extent does storytelling promise relief for those toiling in the lowlands of sustainability communication, or is it just another instrument in the communication toolbox (Janda & Topouzi 2015)?

Ultimately, we think it's a bit of everything. The results of the SusTelling project and the contributions from the practice fields in this volume trace not only the growing importance of the tool "storytelling" but also the manifold challenges in the science and practice of Storytelling for Sustainability (SusTelling). Additionally, this volume holds a wealth of concrete examples and recommendations for the use of SusTelling approaches in different fields of practice, showing where its potential and limitations are.

In this final chapter, we will first summarize the insights from the individual chapters concerning the overarching questions of the volume. We then condense these insights into three theses. We have chosen this format to emphasize two things: first, these insights reflect our positions as editors, which are not necessarily those of the contributing authors. Second, we are putting up for discussion here recommendations and reflections that we hope, looking forward, will provide an impetus for further engagement with SusTelling in research and practice.

DOI: 10.4324/9781003326144-14

Looking Back

Let us take stock of what we have learned relating to the three broad questions raised in the introduction: What is our understanding of sustainability-related storytelling? What are the effects of sustainability-related storytelling? How can sustainability-related storytelling be taught, learned, and practiced?

Understandings of Storytelling in Sustainability Communication

The first part of this volume unfolds an understanding of SusTelling characterized by three components: the intrinsic characteristics of the story itself, the contexts in which it is told, and the intended effects to be achieved through its telling. We argue here that it is the intended effects in particular that make storytelling for sustainability special and distinguish it from other forms of storytelling. These effects relate to the dual ambition of educating audiences and of supporting change for sustainable development. With its focus on education and reflexivity, SusTelling distinguishes itself from understandings of storytelling that see it primarily as a technique to bring about change without the recipients themselves being aware of it. The requirement to be educational is thus related to an understanding of education as emancipatory, as discussed in education for sustainable development (Michelsen & Fischer 2017). On the other hand, the requirement for SusTelling to also aim at having a concrete impact on sustainability is related to the general goal of transformative sustainability science to develop solutions to tackle pressing sustainability problems (Miller et al. 2014). This can be understood as a safeguard that prevents SusTelling from communicating sustainability issues in ways that undermine and hinder transformational processes toward sustainable development rather than advance them (e.g., greenwashing).

Both of these requirements, of educating and of supporting change toward sustainable development, set out a framework that provides orientation for the use of SusTelling. However, this framework needs to be further specified and cannot take decisions away from us, for example, the question of where reflexive engagement with the contents of SusTelling ends and merely subliminal processing of the message begins, or the question of which effects of storytelling tend to promote a change toward greater sustainability and which tend to hinder it. Regarding both requirements, it becomes apparent that evaluation criteria are needed that are not a part of storytelling itself but have to be applied from the outside. This is one aspect of the understanding of SusTelling presented in this volume.

The practice-oriented chapters provide details of the specific requirements of SusTelling and illustrate them with concrete examples. In some fields of practice, it becomes clear that the focus is primarily on the educational impact of storytelling. The chapters on higher education (see Chapter 5), science communication (see Chapter 7), and corporate communication (see Chapter 8) highlight the educational potential of SusTelling. This impact is primarily seen in how knowledge of sustainability topics is represented for different target groups. By contrast, in Chapter 6 SusTelling is seen as a way for constructive journalism to develop and promote concrete solutions to sustainability problems (sustainability outcome) and at the same time encourage people to act (educational outcome).

Chapter 5 in this volume provides a further perspective on education and sustainability effects. It presents storytelling as an approach to knowledge production that allows individuals to work through, document, and share their lived experiences of oppression and discrimination. Its goal is to mobilize against oppressive and discriminatory institutions, systems, and practices. The concept of emancipation expressed here is rooted in the educational tradition of liberation pedagogy and its focus on fighting social injustice. These are further important examples of the educational and sustainability effects laid out in our understanding of SusTelling.

Effects of Storytelling

As described in the introduction, the assumptions made about the effects of storytelling in various fields of practice in sustainability communication were a trigger for the SusTelling project and its aspiration to contribute to an evidence-based study of storytelling. If we now summarize the insights found in the various chapters of this volume, we cannot avoid reflecting on the concept of evidence itself. When we talk about knowing whether and how SusTelling works, which concept of knowledge do we orient ourselves toward?

In sustainability research, the question of what constitutes knowledge relevant for sustainability transformations has been the subject of intense and sometimes controversial debate for many years (Brandt et al. 2013, Apetrei et al. 2021). There is now widespread agreement that cooperation between different scientific disciplines is needed to explore complex questions about interactions between human and natural systems (interdisciplinarity). Increasingly, however, the view has become widespread in the field that knowledge is not only available within the various sciences on how these interactions take shape and have an impact, what solutions can look like and how processes of change can be initiated, but that knowledge is also produced and available in other areas of society. In so-called transdisciplinary approaches to research, many fields of sustainability science are working on integrating different forms of knowledge and making them fruitful for the development of sustainability solutions (Lang et al. 2012). In addition to traditional scientific forms of knowledge (e.g., epistemic systems knowledge), these include forms of knowledge rooted in specific contexts (so-called situated knowledge, such as indigenous knowledge or practical, professional, and everyday knowledge) (Caniglia et al. 2021).

With this in mind, when we look at the evidence on effects found in the different chapters of this volume, it should be noted that it builds on different foundations and was obtained using different methods. Established scientific approaches such as the field experiment, qualitative interview studies, and systematic literature reviews were used in the first chapters. In the later chapters, on the other hand, science-based knowledge is also referred to, but this knowledge is placed in the context of practice field-specific and experience-based reflection (situated knowledge).

The empirical findings from the SusTelling project presented in the first part of this volume show that there is no simple answer to the question of whether and how SusTelling works. In the field experiment (see Chapter 2), none of the hypotheses was confirmed. The superiority of the narrative text could not be established, neither in general nor in interaction with personal characteristics (sustainability interest and educational socialization). The interviews conducted with the participants following the experiment clearly show that there are fundamental differences between the university students and the vocational students in the perception of the SusTelling-style text, and these extend to the perceived effect. It seems, for example, that narrative forms of communication should clearly show more skeptical recipients what the intention behind this type of communication is (persuasive context) and that, for these individuals, in particular, a good balance between factual content and narrative form is important. These results also invite us to reconsider how much complexity and what kind of texts certain target groups can be expected to understand. Vocational students in particular seemed to be sensitized to textual material being presented in an oversimplified way. In addition, the results indicate that the identification potential of the protagonists plays a crucial role in the impact of the narrative-based material. Finally, the results highlight the need for further systematic research on the interaction between recipient characteristics and the elements of a story.

The reception typology developed in the explorative study (see Chapter 3) shows that SusTelling, in addition to stimulating learning processes (*explorers*), can also provide impulses for changing one's sustainability behavior (*visionaries*) or invite people to mobilize others or society toward sustainable development (*enlighteners*). As a qualitative study, this reception typology does not have to be statistically representative. However, reconstructing reception processes can help us to understand how SusTelling invites critical engagement with issues of sustainable development (Hepp 1998) and how the approach activates the application or acquisition of certain competencies (Wiek et al. 2011). This includes being able to think through future-oriented solutions and examine them for their practical feasibility. We could also observe how strategic competencies are used when the storytelling material stimulates recipients (*visionaries*) to try out new solutions or develop creative ideas. Normative competencies come into play through the reception of SusTelling, especially when it comes to evaluating current states of society and nature (*enlighteners*). In addition, competencies for interpreting the often complex interrelationships of ecological and social problems are also applied during learning processes triggered by the reception of SusTelling that generate new knowledge and a better understanding of abstract information (*explorers*). Last but not least, SusTelling stimulates exchange about sustainability issues, whereby interpersonal skills can be activated and trained through successful narratives.

That not only the reception of SusTelling but also its production produces effects becomes clear in the conversation in Chapter 4 in this volume: learning SusTelling involves different skills that relate to content, value, and craft. The content level leads to an engagement with sustainability issues, and the value level requires the learner to self-reflect on the desired effects and undesired side effects of sustainability communication. At the craft level, learning SusTelling develops expertise in shaping the various internal characteristics and external contexts of storytelling (see Chapter 1).

In the second part of the volume, various effects of SusTelling are made seen in its practical applications. In the field of higher education (see Chapter 5), the educational potential of SusTelling is discussed, pointing to improved learning effects (understanding, interest, and motivation in sustainability issues as well as the ability to communicate knowledge to a broader public). The field of (internal) corporate communication (see Chapter 8) refers to similar effects. The examples from human resource development and employee training emphasize that SusTelling can be used as a self-empowering tool to support the employee development, impart new knowledge, or sensitize them to hazards in the workplace. In the field of journalism (see Chapter 6), the author emphasizes two effects of SusTelling: first, the ability of storytelling to capture the attention of the audience and so, and second, the possibility of increasing memory recall of a message. She explains where these effects come into play in constructive journalism and where further research needs to be done.

In the field of science communication (see Chapter 7), the authors emphasize that storytelling must first understand a nonspecialist public before it can help them understand and reflect on scientific results. SusTelling thus contributes to a core mission of science communication, namely, to create the foundation for informed decision-making in a democratic society. In this chapter, storytelling is seen as having an educational and emancipatory effect on its recipients (citizens). However, the authors also point to problematic effects when storytelling overly condenses scientific processes and findings, thus strengthening the trend to popularize and eventicize science.

Finally, the chapter on higher education (see Chapter 5) points to a completely different effect of SusTelling, one that is more strongly focused on social change. It is about how *testimonios* can be used to work through and share lived experiences of injustice and suffering as a means of promoting social justice. The authors not only point to rich experiences that have been made with this form of storytelling in various fields but also emphasize that it has only been used in a rudimentary way in the field of sustainability.

Teaching, Learning, and Practicing Storytelling?

A third interest of this volume was to gain clues as to how sustainability-related storytelling can be taught, learned, and practiced. Answers to this question are, of course, not independent of those to the previous questions about understanding

SusTelling and its effects. Nevertheless, it is important to address the question of practical implementation at this point to better understand how general insights can be transferred to concrete action in communication practice.

The empirical studies from the SusTelling project have shown that strategies for turning knowledge about sustainability into stories must take into account the expectations and needs of recipients. The typology derived from the reception data in Chapter 3 provides guidance for the practice of storytelling. Audiences who are open to issues related to sustainability generally respond positively to the story format, while recipients with less affinity for sustainability tend to want a balance between factual information and narrative techniques and expect practical recommendations. Balancing factual and narrative content is particularly important if recipients are to find stories credible and have a positive perception of them. On the other hand, storytelling should avoid bold and one-sided representations that offer only normative understandings of an issue. The statements of the students participating in the reception study indicate that the positive influence of SusTelling is limited when topics and information are presented in a too one-sided manner. Recipients feel they are being influenced in a way that deprives them of the opportunity to develop their views or interpretations. Due to the homogeneous composition of the target group, it was not possible to investigate whether this critical reflection practice is also found among other groups with different levels of education.

A more comprehensive view of different objectives in the use of SusTelling is found in Chapters 5 and 8 in the practical fields of teaching in higher education as well as in employee training in corporate contexts. In higher education (see Chapter 5), the authors distinguish between three uses of SusTelling: as a method for teaching–learning, for acquiring knowledge, and for external communication. In corporate communication (see Chapter 8), the focus is also on goals in teaching and learning but primarily emphasizes motivational effects, which are achieved when the gap between knowledge transfer and entertainment value is bridged. For each of these different applications of SusTelling, concrete examples are given in the two chapters. Thus, in the business practice field, SusTelling is shown to support employee learning by arousing their interest in "dry" teaching–learning settings with mandatory participation.

Finally, the conversation between teachers of storytelling and the SusTelling project team (see Chapter 4) summarizes the particular challenges that SusTelling faces on a content, value, and craft level for teaching. The conversation makes clear that learning challenges are not only related to the storytelling skills of the learners themselves but also important to be able to strategically counter external resistance and biases that reject storytelling as an approach that is inappropriate, unsuitable, or even suspicious in a specific field of practice. The "storytelling crisis" in journalism triggered by the "Relotius affair" is cited as an example in this context.

Now that we have brought together the insights from the chapters in this volume and put them in the context of the project's goals, we want to zoom out again and venture an outlook on the future.

An Outlook With Three Theses

Henri Nannen, former editor-in-chief of the German news magazine *Stern*, is reported to have repeatedly demanded of his editorial staff that they should focus on the most important insight or core message of their work as what will later be passed on in the briefest form by readers (tell-a-friend). In German, this principle has also become known as "Küchenzuruf" (kitchen call), going back to a memory of Nannen where his grandfather after reading an article in the newspaper shouts to his wife in the kitchen: "Grandma, they're going to raise taxes again", as the shortest summary and key essence of what he's read (Gerharz 2021). Nannen's demand was made in the field of journalism, where comparatively short texts are produced emphasizing the news value of a topic. Nevertheless, in this spirit, we want to ask ourselves, by taking a step back, what essential insights can be taken away for further work on SusTelling in communication practice and research. From our perspective as editors, we have chosen three theses, each relating to a more conceptual, empirical, or programmatic level.

Thesis 1: Storytelling Can Support the Goals of Sustainability Communication in a Variety of Ways and Should not Be Reduced to Persuasion

A major premise of this volume was that the idea of sustainability makes demands on storytelling both to be educationally effective and to promote social change toward sustainability. With this comes the need to motivate people to change their behavior, influence their perception of certain environmental risks, and present information to them as more credible and persuasive, as recent studies in the field of climate change communication have shown (Gustafson et al. 2020; Morris et al. 2019; Sangalan & Bloomfield 2018).

Undoubtedly, such use of SusTelling would help make the world more sustainable. However, a key message of this volume is that the use of storytelling in sustainability communication should not be limited to bringing about those changes intended by storytelling producers. We say this because, first, it is an empirically open question to what extent such changes in the audience actually result in more sustainable behavior and – if it is the case – whether it can also be sustained over the long term (e.g., the studies on climate change communication discussed in this volume only measure immediate changes and do not make any claims about long-term effects). Second, such a narrow focus is also problematic because sustainable development itself requires deliberation and social learning processes. Although there is sufficient consensus on many points about what needs to be done (e.g., that a plant-based diet is more climate-friendly), it is precisely those problematic situations without simple solutions that need to be widely disseminated through storytelling. Think, for example, of the question of what economic systems should look like in the future and how we can bring them about so that the carrying capacity limits of our planet are respected and a good life is possible for all people.

Such "wicked problems" require searching for options, weighing goals against each other, and developing context-bound solutions that need to be repeatedly reviewed, adapted, and reworked.

Sustainability communication therefore always has the dual objective of being educationally effective and promoting social change toward sustainability. The various contributions in this volume show that storytelling can support these objectives and both stimulate concrete change (communication *of* sustainability) and encourage reflection and proactive thinking (communication *about* sustainability) (Newig et al. 2013).

Thesis 2: SusTelling Can Be Educational if Learning Is not at the Expense of Entertainment and the Story Allows the Audience to Identify With Its Protagonists

In sustainability communication, storytelling has to fulfill some special requirements. Where there are no proven solutions for a problem, complexity can only be reduced to a limited extent. If SusTelling wants to have an educational effect here and make complex topics (more) understandable, it is therefore always faced with the task of not overwhelming the audience with one-sided representations and preventing them from developing their own stance and position (see Wehling 1977). This can be prevented by also considering competing viewpoints, perspectives, and possible solutions. This is a tall order for storytelling – and one that can easily conflict with the nature of storytelling itself. Indeed, the empirical results of the SusTelling project show the importance of stories, first of all, giving the audience the feeling they are immersed in the narrative – or, as communication theory puts it, narrative transportation (Moyer-Gusé 2008; Slater & Rouner 2002). When stories no longer entertain and inspire because their instructive intent is pervasive, the mechanisms of narrative persuasion can no longer work either. To meet these challenges, ambiguity can be incorporated into storytelling – in the form of plot, character, or tonality. One example of such a strategy is the use of hidden object pictures in visual storytelling, which open up the potential for different target groups to identify with different characters and actions and allow different readings through nonverbal images.

Thesis 3: Not Only Can Sustainability Communication Use Storytelling, but It Can Also Generate New Forms of Storytelling

The few scientific studies on how storytelling can be used in climate change communication often focus on examining research results on storytelling in other fields and deciding whether they apply to a specific sustainability topic. It is undoubtedly right and important to build on the current state of research and make innovations from other fields fruitful for sustainability communication. Barth and Michelsen (2013) speak here of "outside-in" innovations, that is, innovations imported into one's field from outside. The contributions in this volume show that an evidence-based approach to storytelling does not lead to greater clarity. The question of which goals are achieved by which type of storytelling for which audience in which way cannot be expressed in a recipe. The experimental findings of the SusTelling project show that storytelling is not superior to other approaches to communication per se. At the same time, the qualitative findings open up perspectives on why this is so by uncovering different requirements and modes of reception in sustainability-related storytelling.

Through these empirical, conceptual, and synthesizing studies on what a story is, what counts as narrative, and what the particular requirements are for storytelling in a sustainability context, this volume has sharpened the contours of SusTelling as a field at the intersection of storytelling and sustainability communication. This also raises the question of how SusTelling can and must evolve beyond merely importing existing approaches to storytelling and become more effective in bringing about socio-ecological change. How can SusTelling go beyond shallow communication and have disruptive effects (in a constructivist sense) that challenge people to engage with their values and attitudes? How can multiple perspectives and identities be combined in sustainability stories so that they are supporting pillars, instead of obtrusive elements, of the narrative and become entertainingly educational? And what might activist forms of SusTelling look like that support sustainable change through a narrative reworking of lived experiences in the face of daunting obstacles? Future conceptual and empirical research on these and other questions could help not only to increase the effectiveness of SusTelling but also to export – in the sense of inside-out innovation – new formats of SusTelling to other fields.

References

Apetrei, C. I., Caniglia, G., Wehrden, H. von, & Lang, D. J. (2021). Just another buzzword? A systematic literature review of knowledge-related concepts in sustainability science. Global Environmental Change - *Human and Policy Dimensions, 68*, 102222.

Barth, M., & Michelsen, G. (2013). Learning for change: An educational contribution to sustainability science. *Sustainability Science, 8*(1), pp. 103–119.

Brandt, P., Ernst, A., Gralla, F., Luederitz, C., Lang, D. J., Newig, J., Reinert, F., Abson, D. J., & Wehrden, H. von (2013). A review of transdisciplinary research in sustainability science. *Ecological Economics, 92*, pp. 1–15.

Caniglia, G., Luederitz, C., Wirth, T. von, Fazey, I., Martín-López, B., Hondrila, K., König, A., Wehrden, H. von, Schäpke, N.A., Laubichler, M.D., & Lang, D.J. (2021). A pluralistic and integrated approach to action-oriented knowledge for sustainability. *Nature Sustainability, 4*(2), pp. 93–100.

Gerharz, M. (2021). *The Tell-A-Friend Impulse.* [https://bit.ly/38FtMXK]. Accessed on May 20, 2022.

Gustafson, A., Ballew, M. T., Goldberg, M. H., Cutler, M. J., Rosenthal, S. A., & Leiserowitz, A. (2020). Personal stories can shift climate change beliefs and risk perceptions: The mediating role of emotion. *Communication Reports, 33*(3), pp. 1–15.

Hepp, A. (1998). *Fernsehaneignung und Alltagsgespräche* [Television appropriation and everyday conversations]. Wiesbaden: VS Verlag für Sozialwissenschaften.

Janda, K. B., & Topouzi, M. (2015). Telling tales: Using stories to remake energy policy. *Building Research & Information, 43*(4), pp. 516–533.

Kent, M. L. (2015). The power of storytelling in public relations. Introducing the 20 master plots. *Public Relations Review, 41*(4), pp. 480–489.

Lang, D. J., Wiek, A., Bergmann, M., Stauffacher, M., Martens, P., Moll, P., Swilling, M., & Thomas, C. J. (2012). Transdisciplinary research in sustainability science: Practice, principles, and challenges. *Sustainability Science,* 7(Suppl. 1), pp. 25–43.

Michelsen, G., & Fischer, D. (2017). Sustainability and Education. In M. v. Hauff & C. Kuhnke (eds.), *Sustainable Development Policy: A European Perspective* (pp. 135–158). London: Routledge.

Miller, T. R., Wiek, A., Sarewitz, D., Robinson, J., Olsson, L., Kriebel, D., & Loorbach, D. (2014). The future of sustainability science: A solutions-oriented research agenda. *Sustainability Science, 9*(2), pp. 239–246.

Morris, B. S., Chrysochou, P., Christensen, J. D., Orquin, J. L., Barraza, J., Zak, P. J., & Mitkidis, P. (2019). Stories vs. facts: Triggering emotion and action-taking on climate change. *Climatic Change, 154*(1–2), pp. 19–36.

Moyer-Gusé, E. (2008). Toward a theory of entertainment persuasion: Explaining the persuasive effects of entertainment-education messages. *Communication Theory, 18*(3), pp. 407–425.

Newig, J., Schulz, D., Fischer, D., Hetze, K., Laws, N., Lüdecke, G., & Rieckmann, M. (2013). Communication regarding sustainability: Conceptual perspectives and exploration of societal subsystems. *Sustainability, 5*(7), pp. 2976–2990.

Sangalang, A., & Bloomfield, E. F. (2018). Mother goose and mother nature: Designing stories to communicate information about climate change. *Communication Studies, 69*(5), pp. 583–604.

Slater, M. D., & Rouner, D. (2002). Entertainment-education and elaboration likelihood: Understanding the processing of narrative persuasion. *Communication Theory, 12*(2), pp. 173–191.

Wehling, H.-G. (1977). Beutelsbach Consensus (transl. by R.L. Cope). In S. Schiele, H. Schneider, & K. G. Fischer (eds.), *Das Konsensproblem in der politischen Bildung* [The Consensus Problem in Political Education] (pp. 179–180). Stuttgart: Klett.

Wiek, A., Withycombe, L., & Redman, C.L. (2011). Key competencies in sustainability: A reference framework for academic program development. *Sustainability Science, 6*(2), pp. 203–218.

INDEX

Note: Page numbers in *italics* indicate a figure and page numbers in **bold** indicate a table on the corresponding page. Page numbers followed by 'n' indicate a note.

For Product Safety Concerns and Information please contact our EU
representative GPSR@taylorandfrancis.com
Taylor & Francis Verlag GmbH, Kaufingerstraße 24, 80331 München, Germany

www.ingramcontent.com/pod-product-compliance
Ingram Content Group UK Ltd.
Pitfield, Milton Keynes, MK11 3LW, UK
UKHW022314250926
475702UK00046B/785
9781032352695